Predicting Leader Survival in Covert Operations from Congo to Cuba

Predicting Leader Survival in Covert Operations from Congo to Cuba

Joy S. Patton

LEXINGTON BOOKS
Lanham • Boulder • New York • London

Published by Lexington Books
An imprint of The Rowman & Littlefield Publishing Group, Inc.
4501 Forbes Boulevard, Suite 200, Lanham, Maryland 20706
www.rowman.com

86-90 Paul Street, London EC2A 4NE

British Library Cataloguing in Publication Information Available

Library of Congress Cataloging-in-Publication Data

Names: Patton, Joy S., author.
Title: Predicting leader survival in covert operations from Congo to Cuba / Joy S. Patton.
Description: Lanham : Lexington Books, [2022] | Includes bibliographical references and index.
Identifiers: LCCN 2021054969 (print) | LCCN 2021054970 (ebook) | ISBN 9781793641717 (cloth) | ISBN 9781793641724 (ebook)
Subjects: LCSH: Low-intensity conflicts (Military science)—United States—History—20th century. | Heads of state—Case studies. | Espionage—United States—Case studies. | Regime change—Case studies. | Coups d'état—Case studies. | Cold War.
Classification: LCC U240 .P386 2022 (print) | LCC U240 (ebook) | DDC 355.02/180973—dc23/eng/20211115
LC record available at https://lccn.loc.gov/2021054969
LC ebook record available at https://lccn.loc.gov/2021054970

Contents

Foreword vii

Acknowledgments xiii

1 Methodology 1

2 Tehran, Iran, 1952–1953 17

3 Guatemala, 1953–1954 33

4 Indonesia, 1955–1958 47

5 Dominican Republic, 1959–1961 61

6 Leopoldville (Kinshasa), Democratic Republic of the Congo, 1960–1961 71

7 Cuba, 1961–1962 83

8 Republic of Vietnam, May–November 1963 93

9 Chile, 1970–1973 105

10 Political Systems: Regime Change and Leader Survivability 115

11 Conclusion 121

Bibliography 139

Index 171

About the Author 175

Foreword

This book explores the unique relationship between covert action and leader survivability, in particular, how leadership styles and personality traits influence this relationship. The life of a ruler is ephemeral. For those who are lucky, their exit from office is through retirement or old age. For most, their tenure is short, often ending through violent means. The overthrow of rulers by their rivals is a common theme throughout world history, and the strategy remains a popular choice in contemporary warfare. However, despite their frequency, very little is discussed in international relations about covert regime change and its effects on leader survival. The research methodology of fuzzy-set qualitative comparative analysis (fsQCA) will be used to empirically test the hypothesis that the extent of a leader's survivability following a US covert action event is influenced by his or her respective personality trait and leadership style. The dependent variable is survival—whether leaders live or die—and the independent variables are leader characteristics, public dissent, global instability, internal conflict, and regime type. Case studies focus on eight covert action events that were both successful and unsuccessful, including Iran, Guatemala, Indonesia, Dominican Republic, Democratic Republic of the Congo, Cuba, Republic of Vietnam, and Chile in order to answer the question: Can leader survival be predicted based on analyses of US-sanctioned covert action events during the Cold War era?

In *The Guns of August*, Barbara Tuchman famously wrote, "We were not going to misjudge or precipitously push our adversaries into a course of action that was not intended or anticipated."[1] She was referring specifically to events leading up to the outbreak of World War I in August 1914. Forty-four years later during the Cuban Missile Crisis, American President John F. Kennedy referred to her words to express his intentions of not letting events control crisis, but rather that he would control the events using covert actions.

Enigmatic and shrouded in secrecy, covert actions by states have often been described as the shadowy means of influencing people and events. As John Breen notes in his article titled, *Disciples: The World War II Missions of the CIA Directors Who Fought for Wild Bill Donovan*, "Morally dubious as it may sometimes be, Presidents rely on covert action as a vital means by which to implement identifiable foreign policy objectives in support of US national security" (Breen 2016, 71). During the Cold War, and in a post–Cold War world where asymmetric threats regularly evolved rapidly, the "quiet option" was often perceived as an appealing alternative to overt war, but do leaders of covert regime change survive to tell the tale?

Covert actions are not only frequent, they also have broad and long-lasting effects for all states involved. For target states, the consequences are often catastrophic. Moreover, covert operations have brought brutal dictators to power and also fueled and extended bloody civil wars by providing arms and resources to insurgent groups and warlords who could not have continued fighting without external support. Even nonviolent covert operations undermine the target state's self-determination and sovereignty. For intervening states, the consequences can also be costly. Poorly executed operations spark long, costly wars where there would otherwise likely have been a quick and easy victory. Exposed covert operations can provoke an international backlash, one that weakens American diplomatic and political authority. Domestically, these missions can undermine a state's democratic ideals by creating a situation where political leaders are unaccountable to their citizens. Yet, at the same time, foreign-imposed regime changes can also have positive effects. For the intervening state, successful covert operations can bolster their security by replacing a hostile regime with a friendly one or by preventing a neutral state from joining an opposing alliance. More importantly, "To the extent that leaders of foreign states matter for foreign policy, we should care whether one is facing a risk of irregular and unexpected removal" (Ward and Beger 2017, 142).

One of the most notable examples of foreign-imposed regime changes and the positive effects they bring forth include the United States' covert operations to curtail Soviet influence in Eastern Europe by methodically dismantling regimes that Moscow had created since World War II. To achieve its objective, the United States pursued a myriad of covert action tactics to include economic warfare—strategic embargo of key commodities to the Soviet Bloc—as well as psychological warfare in an attempt to influence attitudes in foreign countries. The United States also developed underground resistance movements to facilitate, support, arm, and mobilize guerilla forces to overthrow Soviet-backed regimes.[2] In her doctoral dissertation (now published into a book) titled, *Secrecy and Security: U.S.-Orchestrated Regime*

Change during the Cold War, Lindsey O'Rourke notes, "America's first covert offensive operations sought to roll back Soviet influence in Eastern Europe by toppling the regimes that the USSR had established following WWII" (O'Rourke 2013, 104). Internal divisions within the Soviet Union and its draconian strategy to maintain control through collectivization and purges, along with divisions within the Communist Party, prevented it from initiating a cohesive strategic policy. More importantly, it allowed the United States to exploit its vulnerabilities through economic and psychological warfare.

THE ROLE OF CONGRESSIONAL OVERSIGHT

The 1975 Senate inquiry into possible and questionable activities conducted by the CIA and other intelligence agencies painted an image of a "rogue CIA" that "was conducting covert operations of its own making irrespective of US policy" (Church 1975). Historically, the application of covert political and military actions has been accepted as means of influencing the actions and intentions of a foreign power and achieving military objectives of the sponsor in the target country. Daugherty asserts that "covert action has its place in American foreign policy and Presidential statecraft, empirically proven by the fact that all post-World War II Presidents have relied on it" (2004, 19). His work titled, *Approval and Review of Covert Programs since Reagan*, reinforces the idea that covert action programs are "very much the President's programs: only he approves them" (Daugherty 2004, 62). Garner and Schumacher's primary thesis is based on the concepts of morals and ethics in covert action. In their paper, they advocate for a codified ethical guidance accomplished through ethical training and education. This case study highlights the importance of leader characteristics as well as the power of public dissent in changing the course of Dominican history. Moreover, the United States' nonintervention policy, known as the "Good Neighbor Policy," allowed a dictator to wield unchecked power in his country without retribution. Based on the study's model, the case of the Dominican Republic incorporates leader characteristics, public dissent, and global instability as the key factors that influenced leader survival.

THE CHURCH COMMITTEE HEARINGS

The Cold War provided fertile ground for the use of covert action as the United States attempted to gain and maintain the upper hand against the threats to its interests and those of its allies posed by the Soviet Union.

The overthrow of the Shah Mohammad Reza Pahlavi of Iran in August 1953, the ouster of President Jacobo Arbenz in Guatemala in June 1954, and the attempt to reverse the Sukarno government in Indonesia in 1956–1957 are all examples of the valuable lessons learned from covert action to effect regime change. Daugherty adds, "The failure of the Indonesian program in no way diminished the Eisenhower administration's zeal for regime reversal; it was a valuable lesson ignored" (2004, 141–44). Additionally, on the covert action ladder, propaganda is listed as being low in its relationship to violence and high in its relationship to plausible deniability. On the other hand, paramilitary force is ranked as high in its relationship to violence and low in its relationship to plausible deniability. In essence, covert action is an effective tool of US foreign policy. Moreover, it will always exist and is an effective tool of American foreign policy only if leaders understand its benefits, limitations, and risks of long-standing consequences. The post–Cold War world has consistently provided a new set of challenges, and presidents often must rely on covert action to counter global threats. In the hostile environment of today's global stage, hostile nations threaten America's national security and transnational terrorism, or non-state actors exploit ungoverned space to launch asymmetric attacks for exponential effects.

As Lowenthal notes, "covert actions are extraordinary steps, something between the states of peace and war" (2009, 167). Despite the lengthy oversight mechanisms, elaborate approval and review processes, meticulous Congressional oversight measures, accountability, and budgetary considerations, the question of morality remains the crux of what makes covert action so controversial. Johnson defines covert action as "a phrase used to identify the pursuit of American foreign policy objectives through secret intervention into the affairs of other nations" (1989, 81). He adds that the appeal of using covert action is evident in the magnitude of funding made available for operations from 1947 to 1986 and that these fluctuations reveal the "attractiveness to policymakers of using the quiet option as an instrument of foreign affairs" (Johnson 1989, 82). During the Gerald R. Ford and James E. Carter presidential administrations, the approval and review process for covert operations evolved into a complicated web of oversight. Following the 1986–1987 Iran-Contra affairs, Congress implemented tougher procedures governing intelligence oversight actions.

While various definitions of covert action and the legal, moral, and ethical considerations of its use have been presented, this book focuses on the significance of leadership traits in determining the probability of leader survival following a US covert action event. Specifically, this section of the literature review is conducted in two phases: first, the researcher searched for the phrase "long-duration leadership," which resulted in 5,082 scholarly sources in the ProQuest and JSTOR research electronic databases. This search led the researcher to consider other search terms such as “irregular leadership,” “leadership duration,” “leadership succession,” and “leader decapitation.” Finally, the researcher considered the terms “personality” and “leadership relationships” in order to infuse a human element into the study. Previous scholarly works on covert action tend only to focus on Cold War case studies without delving into the lives of the leaders who were influenced by such events. In addition to the electronic search, the author also explored the National Archives database to search for leader case studies encompassing the Cold War timeline. After an exhaustive search of scholarly sources and studies, it was apparent to the researcher that, while topics on political leadership and leader duration were popular, integrating these terms into the context of covert action proved to be more of a challenge.

NOTES

1. Every American president has used some form of covert action to control the course of events. This book will focus on the effects of covert action on leader survival.
2. National Security Directive 5412/2 (December 28, 1955).

Acknowledgments

I would like to thank my family and friends, Dr. Robert (Bob) Pauly Jr., Dr. Joseph (J.J.) St. Marie, Dr. Tom Lansford, Dr. Loch Johnson, and Ms. Robin Kelley, editor extraordinaire.

Chapter One

Methodology

The goal of this study is to successfully examine the relationship between leader survival and US-sanctioned covert action in eight countries: Iran, Guatemala, Indonesia, Dominican Republic, Democratic Republic of the Congo, Cuba, Republic of Vietnam, and Chile. In addition, the research will integrate the elegant theory of neoclassical realism to interpret the course of unrest in each of the historic events. The objective is to demonstrate how different factors influence leader survival in covert action events and, ultimately, to test the theory that decision-making and policy are shaped by the outcome of these events. The more aspirational goal of this study is to build a convincing case that leader survival is an important consideration to create a covert action/leader survival model that is generalizable and usable for military planners and decision-makers.

DATA COLLECTION METHODS

Data collection for this project relied heavily on archival sources for primary source documents and existing scholarly, peer-reviewed books on each of the case studies for secondary sources. Primary source documents included detailed memorandums, reports, and biographies of leaders as well as historical and military records. These documents were obtained from archival sources such as, but not limited to, the United States and National Archives, study research archives at Georgetown University, George Washington University, and American University as well as declassified documents and archives offered by the CIA. Each of the selected cases has a large archival and historical record associated with it, thereby aiding in making an in-depth data collection effort possible. Existing research by other scholars was used as a guide to locate pertinent records. This method also served as a check on the potential bias of secondary sources used in this research.

METHODS FOR DATA ANALYSIS AND SYNTHESIS

To filter through vast amounts of information in order to get to the essence, it is important to know what data to find, how to use it, and how to organize it once it is acquired. Specific questions may be useful in gleaning information from vast swaths of information. A full discussion of the reasoning for selections of degree of membership with each variable will be conducted after the presentation of each case study. This project was suitable for a qualitative comparative analysis (QCA) research design because the variables presented were interchangeable and could be mixed and matched according to the researcher's needs.

In their book titled, *Case Study Methods in the International Relations Subfield*, Bennett and Elman note that "the prominence of qualitative methods in IR thus reflects these methods' advantages in studying complex and relatively unstructured and infrequent phenomena that lie at the heart of the subfield" (2007, 171). They add that "the study of IR will undoubtedly progress most rapidly through a combination of methodological approaches" (2007, 171). Bennett and Elman conclude that "Ragin's QCA/fuzzy-set approach has sparked a considerable debate among qualitative and other methodologists and some interest in applying both the crisp and/or fuzzy-set versions in the study of IR" (2007, 182). They add that the use of QCA has not been as prevalent in the study of IR as it has been in comparative politics. However, "We anticipate that QCA/fuzzy-set is likely to become significantly more prominent over time" (Bennett and Elman 2007, 182).

SURVIVAL: THE DEPENDENT VARIABLE

The focus of this project is on leader survival following a covert action event. Therefore, the dependent variable used in this project to determine whether a leader's personality trait and leadership style or other factors influenced his fate following a covert action event.

CAUSAL CONDITIONS OF LEADER SURVIVAL

This project sought to test the effects of five causal conditions on the fate of leaders. To be more specific, these causal conditions are leadership characteristics, public dissent, global instability, internal conflict, and regime type. In keeping with the QCA method, the following cases use existing definitions to describe each casual condition and establish attributes for use in assigning degrees of membership for each of the cases with the causal conditions. The intent of this process was not to establish specific cut points between each gradation in degree of membership, but to establish characteristics of each factor that assists the research in assigning the degree of membership of each case with the causal factor while taking into consideration the within-case context of each case that influenced the individual degree of membership.

Leader Characteristics

Leader characteristics can be addressed by asking the questions:

1. What is the ethnicity of the leader? (Is he a member of minority population?)
2. Does the leader have a college degree? (If so, is it from within his country, or was he educated abroad?)
3. Is the leader over the age of forty?
4. What occupation did the leader hold prior to taking office?
5. What personal experiences shaped his personality?
6. Did he experience trauma in his childhood that might have shaped the leader he became later?
7. How did his personal relationships influence his leadership style? Did these relationships impede or enhance his success?

Public Dissent

Public dissent can be addressed by examining the following questions:

1. Did the public support this leader or not?
2. Did he assume power during a time of peace?

Global Instability

Global instability can be addressed by asking the questions:

1. Was there a major war or conflict occurring at the time the leader was deposed?
2. Were any conflicts occurring in bordering countries at the time the leader was deposed?

Internal Conflict

Internal conflict can be assessed by asking the following questions:

1. Was the country in a state of crisis at the time of the covert action event?
2. Had there been a coup, revolution, or regime change within five years prior?

Regime Type

Political institutions can be "autocratic; democratic; not independent; occupied by foreign troops; ruled by a provisional government charged with overseeing a transition to democracy; or lacking a central government" (Geddes 2014, 313). Regime type can be assessed by asking:

1. Was this a democratic society at the time of the covert action event?
2. Was this a totalitarian regime?

ENSURING TRUSTWORTHINESS

In this research, only archival documents are explored. As such, there is not a requirement for creating the same context and environment with human participants. If the archival documents are limiting, then interviews with participants would be used as an alternate plan. Additionally, not using human subjects will eliminate the requirement for an institutional review board (IRB). Most contemporary case studies on covert action events have yet to be declassified. However, this book is focused on the Cold War period, which should allow for plenty of opportunities to gather declassified data. Future studies will delve into declassified documents from contemporary conflicts. To mitigate potential issues of trustworthiness, every effort was made in the data collection phase to incorporate multiple perspectives of the covert action case study.

VALIDITY

This study follows the same pattern in the analysis of each case study to ensure trustworthiness and validity. Morse et al. add, "If the principles of qualitative inquiry are followed, the analysis is self-correcting. In other words, qualitative research is iterative rather than linear, so that a good qualitative researcher moves back and forth between design and implementation to ensure congruence among question formulation, literature, recruitment, data collection strategies, and analysis" (2002, 17). Moreover, "data are systematically checked, focus is maintained, and the fit of data and the conceptual work of analysis and interpretation are monitored and confirmed constantly" (2002, 17).

LIMITATIONS OF THE STUDY

The most critical limitation of this study is potential gaps in the historical record of the cases and the risk that many cases still remain classified. Additionally, leader profiles may be difficult to acquire, particularly if they involve intimate details of the individual and associates., Personal preference and subjectivity of the research could bias the analysis of data and formulation of findings. The utilization of quantitative "survival testing" would also be an advantage, but the researcher elected to make this a solely qualitative research design. The utilization of the Cox Hazard Model, a regression analysis statistical technique, would have added an interesting dimension to the study. In fact, there is a whole section of survival analysis research that uses the Cox Hazard Model. Additionally, Cox Models are best used for cases which have identical event times. In this study, events occurred at different times. Finally, the limited number of cases used in this study may limit the possibility of generalizing the findings of this study to the analysis of other covert action research.

Morse et al. (2002) offer a step-by-step approach to verification: "First, the aim of methodological coherence is to ensure congruence between the research question and the components of the method; second, the sample must be appropriate, consisting of participants who best represent or have knowledge of the research topic; third, collecting and analyzing data concurrently forms a mutual interaction between what is known and what one needs to know; and fourth, thinking theoretically is a key component of verification" (Morse et al. 2002, 17). These components are used together to ensure accuracy and transparency in research.

In *Issues of Validity in Qualitative Research*, Kvale (1989) states, "To validate is to investigate, to check, to question, and to theorize; all of these activities are integral components of qualitative inquiry that insure rigor—whether quantitative or qualitative methods are used, rigor is a desired goal that is met through specific verification strategies" (1989). Validity is an integral component of the methodology.

The researcher took extra care to check, question, and ensure that the research question was firmly tied to its theoretical framework. Each case study followed the same pattern, and the same questions were asked without variation. Additionally, this research is the starting point for future projects, which will allow the researcher to investigate specific aspects of the phenomena and to study components of the topic that will eventually contribute to the larger theoretical model.

LEADERSHIP CRITERIA

In this research, the big five traits to leadership model is used as the baseline for distinguishing leadership characteristics and personality traits. Using this model, a combination of causal connections between personality traits and ultimate fate are explored and set within the context of the covert action case studies. For example, "extraversion is more strongly related to leader emergence than leader to leader effectiveness" (Judge et al. 2002, 773). In his journal article titled, *Creativity, Divergent Thinking, and Openness to Experience*, McCrae adds that, "Open individuals may be better leaders because they are more creative and are divergent thinkers, because they are risk takers, or because of their tendencies for esoteric thinking and fantasy" (52, 1987). Many scholars, including de Mesquita, Siverson, and Woller (1992), de Mesquita and Siverson (1995), Goemans (2000), and Chiozza and Goemans (2003), have published works on the impact of interstate war on leader survival. Additionally, books on the decapitation of terrorist leaders and its effects on organizational collapse have shown that "leader capability" and "social network analysis" are two primary reasons why insurgencies endure (Tiernay 2015, 178). Other factors to consider include whether the organization is shaped hierarchically or in a more decentralized manner.

Leader characteristics (LDR) sets the foundation for the case study in relation to the leader and the political and socio-cultural environments in the country at the time as well as any relationships the leader had with American or other mentors, confidantes, decision-makers, or advisors. Details captured in declassified national security documents and psychological profiles of leaders gleaned from CIA documents and personal biographies contributed to the assessment that those who had more agreeable personalities and enjoyed close ties with the United States may have experienced a more favorable fate. Conversely, leaders who did not enjoy close relations with US advisors, who had disagreeable, violent, and irrational temperaments, tended to amplify the political and socio-cultural situation in the country at the time, thereby resulting in a less favorable fate.

Public dissent (DISS) explores the existence of protests or mass mobilization of local populace against the leader at the time. The existence of mass protests or popular dissatisfaction with the leader indicates a weak support network for the individual. Intense tribal rivalries, hatred of minorities, discrimination and violence toward ethnic minorities, and any type of social injustice are fuel for mass social movements against a leader and their administration.

Global instability (INS) indicates that the country endured war or conflict that may have altered its future course and demonstrates the influence of powerful political actors: the United States represented democratic values and freedom while the Soviet Union's communist doctrine espoused socialism at all levels of society. A leader's position in either camp was an indicator of a favorable or unfavorable fate. In some cases, such as that of Indonesia, a leader advocated for democratic values but had intentions to ground it in communist doctrine.

Internal conflict (CON) indicates that the country has experienced or is experiencing a coup and/or revolution. Constant churning of political turmoil weakens the country's foundation and creates chaos at all levels. In addition, internal conflict wreaks havoc on a country's socio-economic conditions and eliminates it from being a competitor in the global economy.

In many of the case studies under consideration in this study, including Iran, Guatemala, Chile, and Democratic Republic of the Congo, the country involved had abundant amounts of natural resources that were seized or exploited by external forces. Actions by leaders to regain these natural resources often led to the leader's unfavorable fate.

Regime type (REG) describes the type of political system in place during the time of the covert action event. Iran was a parliamentary political system operating as a fledgling democracy; Guatemala was a praetorian-centric authoritarian regime that placed a tremendous amount of authority and influence on the military; Indonesia was a communist regime thinly veiled as a pseudo-democracy; Chile had a reputation for being a long-standing democracy until the election of Salvador Allende; Democratic Republic of the Congo was a tribal-based governance that had a passionate desire to become a democratically free country free from colonialism's brutality; Dominican Republic's decades-long reign of terror at the hands of a brutal dictator; the Republic of Vietnam was divided between the communists led by Ho Chi Min in the north and a puppet president reluctant to risk his safety to pursue democracy for his people in the south; and Cuba never pretended to be anything more than it was. Castro lived, breathed, and died a communist leader. Surprisingly, the type of regime had a minimal effect on a leader's favorable or unfavorable fate. Below, Table 1.1 offers a visual summary of the relationships of the variables in relation to the factors of the model and their outcome. It also demonstrates the different combinations that can be used in the QCA methodology.

Table 1.1.

Country	Leader Characteristics (LDR)	Public Dissent (DISS)	Global Instability (INS)	Internal Conflict (CON)	Regime Type (REG)	Outcome
Iran	Mostly In	More In Than Out	Fully In	Fully In	Mostly Out	Success
Guatemala	More Out Than In	More In Than Out	Fully In	Fully In	More Out Than In	Success
Indonesia	Fully In	Mostly In	More In Than Out	Fully In	Mostly Out	Unsuccessful
Dominican Republic	Fully In	Fully In	More In than Out	More Out Than In	Mostly Out	Successful
Democratic Republic of the Congo	Fully In	Mostly Out	More Out Than In	Fully In	Fully In	Successful
Cuba	Fully In	Mostly In	Fully In	More Out Than In	More Out Than In	Unsuccessful
Vietnam	Mostly Out	Mostly In	Mostly In	Mostly In	More Out Than In	Successful
Chile	Fully In	Mostly In	Fully In	Mostly In	More In Than Out	Successful

Most successful covert action events demonstrate a general adherence to the factors of the model and indicate the validity of the model. The exceptions to this observation indicate a potential relationship between the factors of the model, combinations of factors, and a successful outcome. All of the successful cases, except for Guatemala and the Republic of Vietnam, demonstrated a Fully In or Mostly In degree of membership with the factor of leadership characteristics as predicted by the hypothesis. In the successful cases that did not display a strong membership with this factor, they did appear to display a strong adherence to global instability and internal conflict. In addition to indicating that successful cases do not necessarily adhere to all the principles of the model, these cases point to a relationship between leader characteristics, global instability, and internal conflict. Specifically, in cases where leadership characteristics are not as strong, external influences creating global instability, as well as internal conflict, resulted in a successful covert action event. The other general observation from this initial review of the data is the importance of leadership characteristics and their effects on regime type. This result indicates the reasonable assumption that a strong leader personality profile is significant in predicting leader survival.

The case studies presented in this book analyze the criteria of leader survivability following a covert action event by using the QCA method. The key finding could be that charismatic leadership style plays an insignificant role in leader survival. Additionally, charismatic leadership styles may attempt to maximize power sharing among leaders. Also, its significance is diminished depending on the security environment at the time of the event. As Ragin notes, "The use of QCA allows multiple combinations of variables which result in a particular outcome. Its strength is that it allows for the examination of different combinations of variables that are consistent with an observed pattern" (Ragin, 2008). In essence, QCA offers a balance between pure quantitative statistical analysis and qualitative methodology based on case studies, surveys, focus groups, and elite interviews, among others. In his article titled, *Using Case Studies to Expand Economic Models of Civil War*, Sambanis (2004) argues that "by combining statistical and case study work we can better understand the political processes that lead societies to civil war" (Sambanis 2004, 269). Other scholars are following suit by integrating statistical methods into a qualitative template. Similarly, Bennett and Elman conclude, "The benefits of juxtaposing statistical and case study methods are well-illustrated in recent research on civil and ethnic conflicts, including the mechanisms behind individuals' participation in such conflicts and the structural conditions that make violence or reconciliation likely" (2004, 187). QCA studies have flourished in disciplines such as political science, sociology, public administration, and organizational and management studies.

According to Ward and Berger (2017, 143), the ARCHIGOS database classifies irregular exits into ten subcategories that may be used to predict leader survival.

From the coding manual, but ordered by occurrence:

1. Leader removed by domestic military actors without foreign support.
2. Leader lost power as a result of domestic popular protest without foreign support.
3. Leader removed by domestic rebel forces without foreign support.
4. Leader removed by other domestic government actors without foreign support.
5. Leader removed in a power struggle within military, short of coup, i.e., without changing institutional features such as a military council or junta.
6. Leader removed through assassination by unsupported individual.
7. Leader removed in an irregular manner through other means or processes.
8. Leader removed by domestic rebel forces with foreign support.
9. Leader removed by domestic military actors with foreign support.

10. Leader removed by other domestic government actors with foreign support.

As with other forms of qualitative research, the congruence method is used in tandem with QCA to impose a theory in the examination of imperfect data. As a result, QCA experiences the same difficulties in causal inference as other qualitative and quantitative methods that use non-experimental data (Ragin 2008). In addition, results from the QCA method can be weakened by spurious correlation between variables as a result of excluding other potential causal variables not included in the theory. Therefore, Ragin cautions that adherence to the "principle that 'context matters' is central" to the QCA method (Ragin 2000, 66). Applying specific templates to leaders may help policymakers make better decisions on the utilization of covert action. For example, an event may have resulted in the death of a leader, thereby resulting in a rise in violence and requiring the need for external security from external sources. In other cases, the death of a leader, as a result of covert action, may have had no impact on the local security environment and, as a consequence, had no residual rise in violence. These are just two examples of how treating two similar cases may result in faulty analysis.

According to Ragin, the combination of a congruence method case-oriented approach and QCA should mitigate the weaknesses of both methods. The congruence case study method will allow for the testing of the explanatory value of the model within individual cases while minimizing misrepresenting the conditions of the cases and allowing for the introduction of other causal variables not included in the model that could be overlooked. QCA will allow for the identification of patterns among the cases that are not apparent in the individual cases and minimize any deceptive data present. Taken together, these two methods should provide a better understanding of the complex causality operating during covert action events.

THE RESEARCH SAMPLE

Three main selection criteria were used to select the cases for this research. The first selection criterion is covert action event initiated by the United States, the installation or attempted installation of a new leader by force, and leaders who were in power at the time of the event. Excluded cases included internal intervention by domestic power/recent or ongoing political crises. As a result, eight cases met the selection criteria. Moreover, these cases were drawn from covert action events in Asia, South America, Sub-Saharan

Africa, the Middle East, North Africa, and Europe. The main purpose of this criterion is to facilitate the measurement of impact of covert action events on leader survival. Additionally, the cases selected for analysis have sufficient primary and secondary sources of data that are well-documented and demonstrate international effects and not just those at the local level.

The second selection criterion is the covert action event must be conducted by a foreign force. This criterion was added for two reasons. First, studying cases involving external intervention forces increases the applicability of the findings of this research to future covert action events conducted by the United States and other major powers. Additionally, the results of this book are intended to support military and civilian policymakers in their decision to use covert action in the future. The second reason for this selection criterion is to provide cases for analysis of the personality traits and leadership styles of foreign leaders. It would be less significant to measure the influence of informal local leaders in small-scale interventions than it is to measure the impact of leaders in grander and more sensational covert action events.

The third selection criterion is the exclusion of internal interventions as a result of domestic and civil unrest. While these cases are interesting and may provide good sources for future research, the outcomes of leader survival are best studied in a more confined scope with clearer operating parameters. Assassinations, displacements, and disappearances of leaders occur all over the world. According to Ragin, fuzzy-set QCA (fsQCA) requires settled cases with adequate primary and secondary sources to make informed judgments on the degree of membership cases demonstrated with a model. In the cases of contemporary covert action events in Southwest Asia, much of the relevant data remains classified.

Using these criteria, covert action events taking place in Iraq and Afghanistan from 2001 up to the current date were not selected because these events have yet to be declassified. However, this book may be used as a template for future research when these case studies do become available at the unclassified level.

According to Ragin, "with fuzzy sets it is possible to have the best of both worlds; namely, the precision that is prized by quantitative researchers and the use of substantive knowledge to calibrate measures that is central to qualitative research" (2006, 67). Attention was paid to selecting cases that are conducted by external powers, but with as much variation as possible in the associated independent variables. The wide range of cases covers over forty years of covert action events occurring in varied geographical, social, and political settings, and involving leaders of different backgrounds.

The cases selected for this book are eight well-established, declassified covert action event studies. The size of the sample was based on an initial review of covert action events during the Cold War, and all available cases were included in the sample.

The first case, Iran (1952–1953), involved a coup d'état under a constitutional monarchy with democratically elected Prime Minister Mohammad Mossadegh. The primary objective was a joint British and American effort to orchestrate a coup d'état against Mossadegh in favor of General Fazollah Zahedi and Shah Pahlavi. In addition, the primary propaganda involved funding dissident groups and sabotage. Type of covert operation was preventative in nature with a successful regime change.

The second case, Guatemala (1953–1954), involved a coup d'état and dissidents with a target government of democracy headed by Jacobo Arbenz Guzman as well as a proposed government with a regime led by General Castillo Armas. The coup against Arbenz involved CIA recruited, trained, armed, and funded dissident forces to stage a mock invasion of the country; propaganda; sabotage; regime destabilization; and limited direct American fighting. The type of covert operation was preventative in nature with a successful regime change.

The third case, Indonesia (1955–1958), involved a coup d'état with a target government and an authoritarian regime led by President Sukarno and a proposed government of non-communist, pro-American regime. Support for coup d'état plot against President Sukarno resulted in an unsuccessful regime change.

The fourth case, Dominican Republic (1959–1961), involved coup d'état and assassination with a target government led by Rafael Trujillo and a proposed government of non-communist, pro-American regime. Support for coup d'état and assassination of Trujillo resulted in a successful regime change.

The fifth case, Republic of the Congo (1960–1961), involved a coup d'état and assassination with the target democratic government led by Prime Minister Patrice Lumumba and a proposed government of a non-communist pro-American regime. At least two assassination plots were attempted against democratically elected Prime Minister Patrice. The type of covert operation was preventative in nature with a successful regime change.

The sixth case, Cuba (1961–1962), involved coup d'état and assassination plans with a target communist government led by Fidel Castro and a proposed government of pro-American regime led by José Cardona, who was a Cuban exile in the United States. Support for coup d'état resulted in an unsuccessful regime change.

The seventh case, Republic of Vietnam (1963), involved coup d'état with a target government led by Ngo Dinh Diem and a proposed government of pro-American regime led by General Doung Van Minh. Support for coup d'état, that inadvertently led to the assassination of South Vietnamese ruler, Ngo Dinh Diem resulted in a successful regime change.

The eighth case, Chile (1962), involved a coup d'état with a target government led by democratically elected Salvador Allende and a proposed government of democratically elected Christian Democrats led by Eduardo Fre and a military junta. Support for coup d'état resulted in a successful regime change.

The criteria for successful and unsuccessful covert action events are defined as a leader surviving or not surviving US-sanctioned covert action. These cases provide numerous within-case comparisons with application of the same event, and they are all differentiated by geographic location, social and security environments, and leadership style.

One of the benefits of QCA is that it is a "case-oriented analytic technique that systemically addresses small numbers of cases (i.e., 5–50) by applying Boolean algebra to implement principles of comparison used by scholars engaged in the qualitative study of macro social phenomena" (Ragin 2005, 34). Using the survival of leaders as an example, leader background, security environment, political institution, and personality/trait are necessary conditions that determine a leader's fate following a covert action event. Regression models can only tell the researcher the extent of the frequency of such conditions' absence. QCA treats each variable the same and allows for the flexibility of mixing and matching variables to determine various outcomes. It also pays attention to the presence or absence of certain conditions and their frequency. Additionally, it uses binary, nominal-scale variables (assigned as 1 and 0) in order to ensure accuracy. This study employs the RStudio software program that is commonly used with QCA. The key objective is to use this technique to determine the significant factors that influence leader survival. The limited number of cases and the infinite causal combinations allows the researcher to assess relationships between the independent and dependent variables with a level of detail that does not exist in regression methodology.

According to Ragin, "considering the small number of cases, the appropriate number of variables should be four to six" (1987, 67). Additionally, he notes, "There are two conditions or states in Boolean algebra: true (or present) and false (or absent). These two states are represented in base 2: 1 indicates presence; 2 indicates absence" (1987, 86). This study selects six key independent variables as conditions for outcome. For example, leaders who are characterized as charismatic are assigned 1 and those who are not are assigned 0. Leaders who are age forty or below are assigned "1" and those who are not are assigned "0." Leaders who left office alive are assigned "1" and leaders who did not are assigned "0." Leaders who were democratically elected to office are assigned "1" and leaders who were not are assigned "0." Those who received a graduate degree or higher received a "1" and those who did not received a "0." Those who served in the military received a "1" and those who did not received a "0."

QCA

The use of QCA has become increasingly popular in the field of international relations over the past ten years. While most scholars still prefer to use quantitative data or a mixed methods approach, the QCA method is gaining popular ground, particularly in terms of research papers presented at academic conferences such as those sponsored annually by the International Studies Association as well as the American Political Science Association. An important component of QCA is that the case studies are limited in number, thereby allowing the researcher to focus on the small and intimate details of each historical case study. The goal of this study is to apply the well-established international relations theory of neoclassical realism into this study on leader survival and covert action.

USE OF THE TRUTH TABLE TO REPRESENT DATA

While Boolean algebra is the technique used to make qualitative comparisons, the framework of the analysis is provided by a raw data matrix that is otherwise known as the truth table. According to Ragin, "The idea of the truth table is simple. Once the data have been recoded into nominal-scale variables and represented in binary form (as in 1s and 0s), it is necessary only to sort the data into their different combinations of values on the independent variables" (1987, 86). Additionally, each combination of variables translates to a separate row on the truth table. Once the table is constructed, independent variables are assigned an output value and these values are compared to the dependent variable and summarized through the truth table.

Table 1.2.

Variable/ Membership	*Fully In*	*Mostly In*	*More In Than Out*	*More Out Than In*	*Mostly Out*	*Fully Out*
Leader Characteristics						
Public Dissent						
Global Instability						
Internal Conflict						
Regime Type						

Truth tables may have as many rows as logically possible. Additionally, Ragin reminds the reader that, "Technically, there is no reason to include the frequency of each combination as a part of the truth table; these values are included to remind the reader that each row is not a single case, but more of a summary of all the cases with a certain combination of input values" (Ragin 1987, 87).

Ragin uses the example of collapsed regime to explain Boolean Logic and QCA: "Consider the collapse of military regimes. Assume that there are three general conditions that cause military regimes to fall: sharp conflict with older and younger military officers (A), death of a powerful dictator (B), or CIA dissatisfaction with the regime (C)" (Ragin 1987, 89). He adds that "any of these three conditions may be sufficient to prompt change" (1987, 89).

Table 1.3.

Condition			*Leader Survival*	*Number of Instances*
A	B	C	F	O
0	0	0	0	0

A = Influenced population through charismatic leadership style/personality trait.
B = Entered office through non-democratic means/force/coercion.
C = Corrupt/unstable political institution.

Each of the rows with one causal factor present and two absent can be combined with rows with two causes present and one absent because all these rows have the same outcome (F), and each pair differs in only one causal condition.

CAUSE AND THEORETICAL EFFECT

Ragin notes that "a cause is both necessary and sufficient if it is the only cause that produces an outcome and it is singular" (1987, 99). Using this topic as an example, a leader who is able to influence people through charismatic leadership and personality traits (A), combined with their rule through coercive or forceful means (B), within a corrupt/unstable political environment (C) may increase the likelihood of death following a covert action event. The theory of neoclassical realism is used to demonstrate the importance of the decision-making/policy implications in relation to survival. The theory allows the researcher to analyze leader behavior in order to assess the probability of their survivability following a covert action event. In addition, it supports the argument that a leader is more likely to survive a covert action event if that individual is a confident leader who assumed power through democratic, nonviolent means and is operating in an environment with a stable political structure.

Chapter Two

Tehran, Iran, 1952–1953

This chapter will explore covert action events undertaken by the United States in Iran from 1952 to 1953. This case study will start with a brief overview of the political environment during that time. Additionally, it will discuss leader characteristics of Iranian Prime Minister Mohammad Mossadegh, and his influence on Iranian society. After discussing the overall aftermath of the covert action operation and how American forces used that action to influence the government, the case study will present a detailed discussion of the degree of membership this case demonstrates for each factor presented in this book's Qualitative Comparative Analysis (QCA) model. Finally, this case study concludes with a summary and an application of the findings.

The August 1953 coup facilitated by the CIA was pivotal to setting the stage for the 1979 Iranian Revolution and planted the seeds of discontent for many years to come. The British exploitation of Iran's national resources in the forms of the Anglo-Iranian Oil Corporation and the decades-long cultivation of precious oil would continue to haunt the region for many decades to come. Journalist Stephen Kinzer in his book *All the Shah's Men: An American Coup and the Roots of Middle East Terror* noted, "Iranians will never forget the 1953 US-supported coup that toppled the nationalist, moderate, and democratic government of Dr. Mohammad Mossadegh and ushered in a closed, dictatorial political system. Iranian society lost one of its most important historical opportunities for the establishment of a democracy" (2008, 62).

In 1953, the United States implemented a plan to oust a popular nationalist hero whose visionary goals would have brought the country closer to the United States while ensuring future wealth for the nation. Instead, the leader was deposed and replaced with a dictator whose draconian policies caused suffering to the Iranian people during his twenty-five-year rule. As Kinzer notes, "Today the West finds itself facing a regime in Tehran that embodies threats far more profound than those that it sought to crush in 1953" (2008, 62). The United States was also concerned over the nationalization of Iran's petroleum industry and the possibility of Tehran leaning toward the Soviet Union rather than the United States at a notable time in history.

This case study highlights the importance of leader characteristics as well as the power of public dissent in changing the course of history. Moreover, global instability created by international superpowers such as the United States, Soviet Union, and United Kingdom fostered an environment of mutual distrust. Based on the study's model, the case of Iran incorporates all the facets of leader characteristics, public dissent, global instability, internal conflict, and regime type. The brevity of the 1953 Iran coup, relative to its domestic and international consequence, is a lesson learned on the dangers of external influences on foreign countries. Through the use of historical documents from presidential libraries, this case study will show how all the factors culminated to create the perfect conditions for a US-sanctioned coup attempt against Mossadegh. The lesson learned from history is that poorly executed coups can, more often than not, create the breeding ground for mass dissent and radical extremism.

The case of Iran identifies leader characteristics as a driving force behind survival. In addition to providing an example of the relative importance of survival in the outcome of a covert operation, the case demonstrates the relative importance of each factor presented in this book's model.

BACKGROUND

As one of the oldest civilizations in the world, Persia was well-known for its innovators of art, literature, and poetry. Above all, Persia's leaders valued knowledge and respected those who sought enlightenment and wisdom. Its resplendent history is full of stories of courageous, battle-forged leaders such as Cyrus, Xerxes, and Darius, and its intellectuals made astounding contributions to the world of philosophy and poetry. This richness formed the fabric of nationalism and communal pride. Situated at the strategic crossroads of Central Asia, South Asia, and the Middle East, Iran (formerly Persia) represents a vivid backdrop in history that is as beautiful as the exquisite tiled mosaics that it produces and patriotism and nationalist pride its people embody. Rising majestically from the mountains of southern Iran, the ancient ruins of Persepolis, former capital of Persia, offer the modern-day tourist a glimpse of Iran's long-lost glory; remnants of a magnificent empire that spanned vast territories and was home to a proud nation of people who placed intellect, art, literature, and culture as valued wealth. Situated along critical trade routes and seeming to have an infinite supply of natural resources, Iranian poets, artists, and architects created a beauty unsurpassed by any other empire at that time.

The 1953 Iran coup sparked a cataclysmic explosion that reverberated throughout the developed world. It occurred as a result of internal political pressure and external influence on a fragile country that had already endured many decades of unequal distribution of wealth, rampant corruption, greed, violence, and misfortune. The country's history was so tumultuous that, in its formative years, political leaders rarely survived a year in power. As Efimenco noted in his article, *An Experiment with Civilian Dictatorship in Iran; The Case of Mohammed Mossadegh*, "Modern Iran still reveals a medieval, fragmented social order, in which sharp cleavages exist between the urban and village centers, between the settled communities and the tribal areas, and between the wealthy and the poor" (1955, 390).

The constitutional movement gave Iran a taste of democracy, albeit fleetingly. Whether from foreign intervention or internal domestic power struggles, Iran has never experienced stability because of rigid military dictatorship and factional political parties. "Subsequent political movements produced a type of fractional party organization which centers on personalities rather than on coherent programs of action" (Efimenco 1955, 390). The communist-led Tudeh Party and the Muslim party, combined with resistance from Tehran's political elite population, created an environment that stifled any semblance of progressive thought or reform. Scavenger-like political buzzards seeking power and control combined with economic inequality created a recipe for chaos and spread seeds of conflict that are still felt to this day. Moreover, the political leaders at the time were considered uneducated, inarticulate, and incompetent. The Iranian people, weary of exploitations from the Qajar-then-Pahlavi Dynasties, desperately needed a change in leadership, and their wish was fulfilled, albeit fleetingly.

LEADER CHARACTERISTICS

Leader characteristics play an important part in the study's model as they have a direct influence on public dissent, global instability, internal conflict, and type of regime. This section will analyze Mossadegh's background, political experience, and leader characteristics as well as other significant and influential factors.

Born in 1882 during the era of *belle époque* (era of beautiful times), Mossadegh was described as "a precocious child who grew up to be a peculiar man with a bald head, long, drooping nose, and thin, sensual lips, and he fainted and howled in public" (de Bellaigue 2012,1). The eldest son of two prominent public figures in Iranian political society, he came of age when Iran was experiencing some of its most violent conflicts. As Kinzer notes, "He showed extraordinary emotional qualities. His boundless self-assurance led him to fight fiercely for his principles, but when he found others unreceptive, he would storm off for long periods of brooding silence" (2008, 54). He had a reputation for being a hypochondriac who exited the scene when things did not go his way. According to his biographers, as a student at l'Ecole de Sciences Politiques (the School of Political Sciences) in Paris, Mossadegh suffered from ulcers, hemorrhaging, stomach secretions, and other ailments. As Kinzer notes, "At times he became so passionate while delivering speeches that tears streamed down his cheeks. Sometimes he fainted dead away, as much from emotion as from any physical condition" (2008, 54). At one time, British Prime Minister Sir Winston Churchill called him a lunatic and named him "Mussy Duck" because of Mossadegh's emotional instability. To the proud people of Iran, Mossadegh was part of the drama of faith and, as such, they embraced him as their nationalist leader and visionary.

Mossadegh completed his academic studies in France and Switzerland and returned to Tehran after completion of his law studies. He applied for Swiss citizenship following graduation from law school but changed his mind after discovering that it would take at least ten years to gain citizenship. Mossadegh was only in his forties during the time that he took power, but his bald head and droopy nose made him look decades older than his real age. Writer de Bellaigue notes, "He ran Iran, which is a big and complicated country, wearing a pair of pajamas" (2012, 1). In an environment of rampant political corruption, inexperienced government leaders, and ever-shifting political alignments, Mossadegh transitioned seamlessly into the political arena and into the open and welcoming arms of his Iranian supporters. He is most well-known for his brazen attempt to nationalize Iran's oil through the British-owned Anglo-Iranian Oil Company (AIOC).

De Bellaigue notes, "Mossadegh was remarkable in his region in that he did not feel the need to be cruel. He was not a soldier, deriving prestige from the pips on his shoulders and the shine on his shoes. He did not borrow his authority, like a Mullah, from God" (2012, 2).

Margaret Hermann's study on foreign policy behavior and personality characteristics of political leaders suggest that "aggressive leaders are high in need for power, low in conceptual complexity, distrustful of others, nationalistic, and likely to believe that they have some control over the events in which they are involved" (Hermann 1980, 7). To many Iranians, Mossadegh was an inspirational leader and a national patriot. He was also an intellectual who earned a law degree from the highly prestigious Paris Institute of

Political Studies. On the other hand, proof of his ability to tend to and support the needs of his followers is lacking. Biographers have noted that Mossadegh gave up support to the Majlis (parliament) because he did not have faith in its members' competence or intellectual abilities. He demonstrated an average level of extraversion, and he lacked a certain level of agreeableness (especially when it came to negotiating with British leaders and members of the shah's royal party). Moreover, he had a deep level of consciousness for the country's suffering and for the promise of democratic freedom. However, Mossadegh's emotional stability was questionable due to his outbreaks and mood swings. Moreover, he was open to new experiences as indicated by his decision to leave Iran to attend school in France, eventually earning degrees from foreign institutions, and obtaining citizenship in Switzerland.

Along with nationalizing oil, Mossadegh passed land reforms, introduced social security and rent control, and strengthened the separation of powers between the branches of government. However, Iran was far from becoming a democracy during this time, and it was, instead, a constitutional monarchy with the shah wielding tremendous power to include the right to appoint or dismiss a prime minister.

NEUROTICISM

McCrae and Costa note that "Individuals who score high on measures of neuroticism lack self-confidence and self-esteem" (1991, 227). De Bellaigue adds, "Mossadegh was instrumental in his own downfall. His judgement failed him in the zenith of events, for it is a captain's job to steer his ship into calmer waters and Mossadegh was driven on by his obsession, into the teeth of the storm. The Shah lived to tell the tale, and he had pretensions to captaincy of his own" (2012, 2).

EXTRAVERSION

Extraversion is related to "leader emergence in groups" (Watson and Clark 1997, 767). De Bellaigue asserts,

> Mossadegh was a hypochondriac known for throwing weeping fits and conducting business from his bed. He had no weakness for girls, boys, money, wine, the pipe or Karl Marx. Any of these vices would have made him more understandable . . . but to his western interlocutors he was a riddle. They found him in his camel's wool aba, or cackling on his haunches in bed, or lying low with his

hands fluttering up and down under his neck. For Mossadegh, the nation's oil represented life, hope, freedom. (2012, 2)

From *Foreign Relations*, 1952–1954, Iran, 1951–1954 (Documents 1–375) United States Efforts To Understand Mosadeq, February 1951–February 1952 (Documents 1–66):

Mosadeq is a master of the rhetoric which appeals to Persian listeners. His voice, in every speech, ranges from a slow reasonable tone to shrill accusations. He mixes wit and poetry into debates upon most serious questions. He plays for emotional reactions from his audience rather than reasoned approbation (page 189).

Perhaps the best recent instance when his oratory won antagonistic listeners to his side was in the Majlis on December 11 after an eloquent opposition had heaped vituperation and some very searching criticism upon his Government. The gentle tone with which he pointed out that he, an old and honorable man, had listened with restraint to everything the opposition had to say gave the impression that it was not he but his critics who were irresponsible. He wove into the tapestry of his speech a thread of reason as though he were a father explaining to a little boy the need to fight for independence in an evil world. Gradually, he brought the color of anti-British feelings into the design, moving from reason to emotion almost imperceptibly so that his listeners felt at the end that he had won a victory over national enemies. (Embassy telegrams 2159, 2162, December 12, 1951)

OPENNESS TO EXPERIENCE

McRae (1987) notes, openness to experience correlates with divergent thinking. Even from the very beginning, the United States was unsure about its approach to Mossadegh, as evidenced by this excerpt from a paper prepared in the Directorate of Plans, Central Intelligence Agency (Central Intelligence Agency, DDO Files, Job 79–01228A, Box 11, Folder 14, Iran 1951–1953):

Mossadeq, in spite of his emotional fainting spells, in the Majlis, his long-winded speeches, and his lack of a constructive program, is too widely admired to be the subject of successful attack. The approach should be against his conspicuous followers, to emphasize the idea that they are deceiving and misleading the grand old patriot. Two methods appear possible:

I am confident that the Department will find the enclosed memorandum of considerable interest. The enigmatic Dr. Mosadeq is to exert considerable influence in Iran in the immediate future. The skill with which he conducts the affairs of the government, and the political course which he follows will

> probably determine the future not only of the vitally important oil resources of this country, but perhaps of the country itself.
>
> Mosadeq is a dramatic demagogue who appears to be without particular wisdom or background for government. These very factors make his presence at the head of the government especially important. He has qualities which, while not commending him to the world, may establish his place firmly in the minds of the people as a popular leader.
>
> Mohammad Mossadeq, Iran's new Prime Minister, is an extreme nationalist. He will attempt to curtail severely foreign influence in Iran and to adopt a neutralist policy toward the East-West struggle. As he is also an impractical visionary and a poor administrator, it is unlikely that he will do very much to solve the country's critical economic and social problems. Nevertheless, because he is an astute politician and has strong popular support on the oil issue at least, he will probably not be easily displaced while that issue is still unsettled.

AGREEABLENESS

Mossadegh was certainly not agreeable to the terms of the British petroleum export policy, and he made it well-known to those around him. In the end, his demise left a legacy of conflict and instability in the region for many decades to come. De Bellaigue writes, "Ousting Mossadegh, in the long run, did great harm to western interests. It was the start of a US policy in support of shoddy Middle Eastern despots, which suffered its first defeat in 1979, when Ayatollah Khomeini's Islamic revolutionaries overthrew the Shah, and unraveled further with the Arab Spring of 2011" (2012, 3).

CONSCIENTIOUSNESS

There is very little evidence to support the relationship between conscientiousness and transformational leadership. However, Mossadegh was very aware of his importance to the Iranian people. After all, he was a patriot at heart and, despite many opportunities to leave his homeland to reside in another country, he remained in Iran.

A dispatch from the Embassy in Iran to the Department of State, No. 878, Tehran, February 16, 1952,

> Mohammad Mosadeq is an aristocratic Moslem Persian, born to wealth, bound by tradition, steeped in classic Persian culture. He is proud to be a Persian and he understands and loves his people. The vices and the virtues he demonstrates are not unusual characteristics in Iran, although he emphasizes those aspects of

> his character which are most useful to his politics." (National Archives, RG 59, Central Files 1950–1954, 788.00/2–1652)

As Kinzer notes,

> Although Mossadegh championed Iranian self-determination, he had little faith in his fellow deputies, and few escaped the lash of his tongue. He accused them of cowardice, of lacking initiative, and worst of all being unpatriotic. His fulminations at the podium were both frightening and theatrical. Gesturing wildly, his hand unconsciously wiping away the famous tears that sprung unbidden from his eyes at times of nervousness or rage, he pilloried his listeners with the righteousness of a priest who suffers with his victims even as he unmasks them . . . Distinguished, highly emotional, and every inch the aristocrat, he believed so totally in his own country that his words reached out and touched the common man. Mossadegh was Iran's first genuinely popular leader, and he knew it. (2008, 58)

Dispatch from the Embassy in Iran to the Department of State:

> To his career, Mohammad Mosadeq brings the most necessary attribute of a demagogue—a dramatic personality. The fainting and the tears which seem so funny to Americans deeply move his Persian listeners. He has a superb sense of timing and of symbolism.
>
> A good demonstration of this quality occurred when he left Tehran in 1951 to present Iran's case in the oil dispute before the United Nations. At the airport, after the dignitaries and a small crowd had arrived, a car drew up some distance from the waiting plane and the limp figure of the Prime Minister was helped by attendants past the crowd. The shrill chanting of the mullahs, the wailing of the crowd, the pathos of the fainting man, who claimed that he would champion his people before the world, were all background to the well-timed moment when the very symbol of Iranian hopes and fears was supported, half-fainting, in the doorway of the aircraft to take a last look upon his people. It was very foolish and unstatesmanlike but very moving. (National Archives, RG 59, Central Files 1950–1954, 788.00/2–1652)

PUBLIC DISSENT

Dissent is an interesting word in the English dictionary—even more compelling when used within the context of a social movement such as the one in Iran in 1953. "Dissent is always dangerous to those who practice it and vexatious to those against whom it is directed. For both the dissenter and their target, dissent stirs up strong emotions and often calls forth strident reactions" (Sarat 2010, 1). Moreover, "Dissenters seek to define and occupy an in-between space, resistant to prevailing orthodoxy but engage with it, nonetheless. It's the ultimate gesture of citizenship. A way of saying: I'm not just passing through, I *live* here" (Gates 1999).

From 1905–1911, the people of Iran experienced a Constitutional Revolution that introduced the concept of democracy and human rights to the populace and advocated for political rights of Iranian citizens against the authority of the shah and members of the Pahlavi Dynasty. During this period, Iran experienced an awakening of a different kind—the discovery, and eventual exploitation, of its natural resources. For Iran, social movements and political dissent were nothing new. Many leaders had been ousted as a result of public revolts. However, in this case, the enemy was not the dictator. Rather, the enemy was the foreign invader and exploiter of Iran's primary natural resource: petroleum. Moreover, Mossadegh had his share of enemies in the form of royalists, anglophiles, defectors, and the Tudeh. The latter, the Tudeh Party, was perceived as growing stronger in strength, momentum, and political cohesion. As Gasiorowski notes, "Traditional Iranian revolts involved the active or passive supporter of all social classes to bring down an unjust arbitrary ruler and replace him with just one" (2004, 3). The effects of the 1953 coup had resonating effects on a society that had long-hoped for democracy and human rights. As Gasiorowski and Byrne note, "Domestically, it brought to an end a vibrant chapter in the history of Iran's nationalist and democratic movements" (2004, 1).

Mossadegh and the Popular Movement were intended to establish a constitutional government as well as democracy and political freedom for the Iranian people. While Mossadegh and the populace believed in these rights, the political machine of Iran was not as liberal. Mossadegh and other Popular Movement leaders believed that as long as external governments owned rights to the country's natural resources, Iran could never truly be democratic or independent.

In a declassified memorandum from Ambassador Henry F. Grady to Mr. Kiplinger from Tehran, July 20, 1951:

> I am hoping the case will continue to build up against Mossadegh to the point where he will have to step down and somebody else is brought in to straighten things out. He is losing support in certain quarters, but it would be a mistake to assume that he has not the Iranian people behind him almost 100%. He has become to the Iranian people a Ghandi. They feel he is fighting for their independence from British domination and they will make any sacrifice to support him.

A declassified telegraph from Tehran to Department of State, No. 165, July 12th, 5 p.m.:

> Embassy has received from several sources increasing indications in political situation that Mossadegh Government may have been considerably weakened by developments of past two weeks. The reasons are: many deputies now realize oil impasse threatens Iran with serious financial difficulties. They state privately government must quickly inform country its plans to avoid or mitigate impending economic crisis; manner in which seized [AIOC] documents were exploited for political purposes by national front members of government has aroused indignation of many, especially those fearing to be implicated, and this controversy has deprived Cabinet of some of its press support; increasing confusion and inefficiency in conduct ordinary government affairs and lack of confidence in integrity of Fatemi; demagogic tactics employed by Mossadegh supporters to keep nationalist sentiment at high pitch have alarmed many deputies, although few have voiced concern publicly; cabinet is becoming increasingly dissatisfied with Mossadegh's penchant for making decisions with a small group of close advisors without full consultation of the Cabinet; Mossadegh has announced group of physicians will submit report on his physical condition. Purpose may be to lay basis for graceful exit.

In the case of Iran, the state's repressive policies and external influences may have channeled oppressive political parties to embrace radical Islam as the state's ideology. The 1953 Iran coup cut short one country's hopes of rising to democratic freedom.

GLOBAL INSTABILITY

In 1951, Iran was a country caught in a vise between global superpowers. The United Kingdom wanted to continue its oil extraction business without disruption, and the United States wanted both countries to find an acceptable middle ground. Mossadegh was known as an outspoken nationalist with an ardent passion for democratic freedoms and human rights. The United Kingdom considered him a nuisance and a potential threat to their plan. President Harry S. Truman had a soft spot for Mossadegh's nationalist ideals, yet Truman's successor, President Dwight D. Eisenhower, and his staff viewed the prime minister's passion as a threat to their very non-communist neighbors.

From a letter written by Mrs. Cornelius Bryce Pinchot to Truman and Secretary of State Dean Acheson:

> The most effective point in the disastrous policy of the Prime Minister, Dr. Mossadegh, is the crushing poverty of the Iranian people. Immediate American aid followed by long-term assistance would answer this point. Many Iranians realize that Mossadegh is leading the country to ruin. Such a move on our part would give them something tangible with which to work. I do not think that it is

> farfetched to suggest that the crisis in Iran has implications that could reach as far as the assassination of the Archduke Franz Ferdinand in Sarajevo in 1914. As things stand now the very best that we can foresee for Iran is rapidly progressing disintegration with little by little the Iron Curtain closing down one of the most strategic nerve centers of the world. The danger of this leading to war is evident. That it could influence the outcome of such a war is also evident.

From a declassified document titled *Iran*, 14 November 1952: "It is of critical importance to the United States that Iran remains an independent and sovereign nation, not dominated by the USSR. Because of its key strategic position, its petroleum resources, its vulnerability to political subversion, Iran must be regarded as a continuing objective of Soviet expansion."

Moreover,

> the loss of Iran by default or by Soviet intervention would:
>
> a. Be a major threat to the security of the entire Middle East, including Pakistan and India;
> b. Permit communist denial to the free world of access to Iranian oil and seriously threaten the loss of other Middle Eastern oil;
> c. Increase the Soviet Union's capability to threaten important United States-United Kingdom lines of communication;
> d. Damage United States prestige in nearby countries and with the exception of Turkey and possibly Pakistan, seriously weaken, if not destroy, their will to resist communist pressures;
> e. Set off a series of military, political and economic developments, the consequences of which would seriously endanger the security interests of the United States. Social unrest is spreading in the wake of nationalist agitation and disruption of the traditional structure of Iranian leadership and institutions.

Iran was a pawn between superpowers on the global stage. On the one hand, British exploitation of its oil within its own country was an impediment to the country's desires to gain democratic freedom and independence. On the other hand, despite its strategic geographic location and abundance of natural resources, its internal politics and corruption prevented it from making a coherent stand against global superpowers.

INTERNAL CONFLICT

This section analyzes the factors that contributed to internal conflict within Iran from 1952 to 1953 and takes a step further to see if these factors led to larger problems in the future. The principal questions to be addressed are two-fold. First, what was the security environment like during this time? Second, did internal conflict result in the spawning of other problems such as state-sponsored terrorism following the 1953 covert action event? From a declassified memorandum written by the White House Chief of Protocol describing the Ambassador of Iran, dated September 24, 1952, it can be reasoned that the security environment in Iran in 1952 was tenuous:

> Telegram No. 189: The Ambassador in Iran (Henderson) to the Department of State Tehran, July 28, 1952—2 p.m.:
>
> During last two days I have recd various hints, including one from son, that Mosadeq was ready for me to call. Accordingly, I visited him yesterday evening. Our conversation, which lasted nearly two and half hours, was both exhausting and depressing. As I listened to him I cld not but be discouraged at thought that person so lacking in stability and clearly dominated by emotions and prejudices shld represent only bulwark left between Iran and communism. As during several previous conversations, I had feeling at times that I was talking with someone not quite sane and that therefore he shld be humored rather than reasoned with. On occasions he resorted to such silly exaggerations and extravagances it seemed almost useless to talk further. At one point I almost decided to abandon our conv when he rptd again and again in monotone that "Iran wld never, never want UK and US to have any differences over it. Iran wld prefer go Communist than cause any trouble between US and UK". There were periods during our talk when he seemed lucid and sensible. Gen impression which he left was however one of deterioration. I have noticed in past that in evenings he is likely to be more tired and to have less control over his emotions. I can only hope his behavior last evening was due to strain of recent events and fatigue and does not indicate serious degeneration. (*Foreign Relations of the United States*, 1952–1954, Iran, 1951–1954, Vol. X. The question of military and economic assistance to Iran; interest of the United States in the settlement of the Anglo-Iranian oil dispute, 1952–1954 [Documents 139–508])

In a telegram from the Secretary of State to the Embassy in Iran, Washington, August 17, 1950—7 p.m.:

> We do not believe present Soviet activities in Iran form part any global policy but are designed solely as part ceaseless attempt by one means or another wean Iran away from west and eventually create new satellite.
>
> Soviet policy towards Iran since end of war has been alternatively hard and soft accompanied by continuous propaganda, border incidents and internal subversion. Assumption arises confusion of Soviet pol is deliberate and designed keep Iran in constant state turmoil with little chance internal stability. Furthermore,

> Iran now has for first time Gov which cannot be intimidated by diplomatic notes and Soviets have probably reached conclusion bullying tactics will not work while ostensibly friendly offer help Iran may, as is certainly possible, be welcomed by many Iranians fearful of future and skeptical of US assistance to such an extent that no Govt eld stay in office which pursued pol openly hostile to Soviets and aligned with US. —Acheson (*Foreign Relations of the United States*, 1950, The Near East, South Asia and Africa, Volume V 611.88/8–1450: Telegram).

Fast-forward to two decades following the shah's overthrow to a new era in history for Iran and the new Islamic Republic of Iran's creation of the Islamic Revolutionary Guard Corp. Its role was to enforce the new country's conservative values. Over the years, this group has spawned independent groups that have been accused of recruiting and training terrorist operatives within and beyond Iran's borders. In the case of Iran, the mistrust stems from great powers such as Russia, Great Britain, and the United States seeking to control and exploit the country and its resources created an unsteady foundation for future diplomatic relations. The US State Department still labels Iran "the most active state sponsor of terrorism" (US Department of State 2002). Islam appears to be the common denominator for most, if not all, of these organizations. Byman (2005) adds, "Iran helped found, organize, and train Hezbollah, eventually creating a strong and relatively independent terrorist group. In exchange, Hezbollah has served Iran loyally, striking Iran's various foreign enemies, helping assassinate Iranian dissidents, and otherwise advancing the interests of the Islamic Republic" (2005, 80).

As Ranstorp (1997) notes, "Iran also gained prestige from supporting Hezbollah. The movement's increasingly successful resistance to Israel raised its stature in the Muslim and Arab world, which in turn reflected well on Tehran. Theology and power politics went hand in hand for Tehran: without being involved against Israel, it was far harder for Iran to portray itself as the revolutionary vanguard of the Muslim world" (Ranstorp 1997). Byman adds, "Hezbollah's capabilities also served as a coercive tool for Iran, enabling it to threaten the United States if it felt Washington's pressure was too strong" (2005, 94). As Kinzer (2009) writes, "Operation AJAX, as the CIA coup against Mossadegh was codenamed, was a great trauma for Iran, the Middle East, and the colonial world. This was the first time the CIA overthrew a foreign government. It set a pattern for years to come and shaped the way millions of people view the United States" (2009, 319).

The internal conflict brought about by intense hatred for external intruders evolved into an anti-American theocracy that ultimately adopted and cultivated terrorism as a tool of statecraft. When diplomatic relations failed time after time, Iran chooses to resort to extreme measures. Terrorist organizations such as Hezbollah, Al-Qaeda, ISIS/ISIL, and those learning new, lethal, and innovative skills in the shadows have confidence that they can rely on Iran to provide them with safe haven, financial, and political support. These events should serve as a stark reminder to any country that seeks to change the course of history by deposing a ruler through violent means. While victors do win coups—such was the case in 1953—they end up losing tragically in the future and paying the price for their so-called victories.

REGIME TYPE

Authors have often referred to Iran as a "regime of religion," but not many know that it sought a regime built on democratic freedom and equal rights. When one hears the word "regime" within the context of Iran, it conjures up a myriad of images ranging from royalty gilded in gold, resting on the Peacock Throne, to charismatic nationalist leaders with a long, drawn face and a crooked nose, to radical Islamic clerks, and to terrorist extremists. The picture is clear: There is no linear way to define Iran and its regime. Despite the pattern of cruel authoritarianism, autocracy, theocracy, pseudo-democracy, repressive and intransigent state, one common denominator exists between all these labels—its defiance against outsiders.

More importantly, the Qajar model of domestic economy relied heavily on exports of textiles and cottage industry products. The allure of European brand of capitalism was becoming a force that was too strong to ignore. In essence, the social fabric of Iran's social class was transforming, and a new, more independent form of government was desired by the people. The bazaar-centered middle class (merchants, artists, etc.) combined forces with the intellectuals to rise against the upper class. The people's attitudes toward the shah were also changing.

Critics and scholars have offered their own interpretation of the 1953 Iran coup. Some scholars assessed that external pressure from Russia, Britain, and the United States created an environment of mutual distrust with Iran. Some viewed Mossadegh as the Middle East's first liberal leader, and others viewed him as an emotionally unstable dictator who initiated programs that demonstrated the pretenses of democracy but, in actuality, were inevitably lost in the greed and corruption of the local and state government. By the time of the coup, the political institution was in shambles and any hope for democracy was fleeting. The only lasting remnant of nationalism and independence is seen in the defiance of the younger generation of Iranians who only know of the 1953 Iran coup from stories told by their parents and grandparents.

CONCLUSION

One of the interesting associations to note in the Iran coup case study is the link between global instability and internal conflict. While the people of Iran had no ill will towards the United States, they did share a bitter resentment towards the United Kingdom. It was not until the time leading up to the coup that the Iranian people started to view the United States in the same category as the United Kingdom. The external influence from these two superpowers had a direct influence on the level of conflict within the country of Iran, during a time when the country was experiencing its own conflict that originated with the encroachment of Islam on its religious culture. While some scholars argue that Iran's fledgling democracy would have certainly failed due to weak political institutions, perhaps it would have taken a charismatic leader with strong patriotism to push the country to the next phase of independence. Mossadegh may have turned the tide of history had he been given the chance to do so.

The US role in the 1953 Iran coup was considered a success by the United States. Moreover, it created a sense of invincibility among US leaders that any country could be influenced by just the right amount of covert force. This case study successfully proved that global instability, combined with internal conflict and leadership characteristics, were more powerful than the other variables such as public dissent and regime change.

Additionally, the United States paid very little attention to Iran's strategic importance in the Middle East. At that time, the focus was primarily on preventing Iran from leaning towards communism. Had Mossadegh emphasized his intellectual strength over his emotional weakness, he may have been able to harness the power of the people into a final push towards democracy.

This case also demonstrated the less influential power of regime type to a leader's survival. Had Iran been on the more extreme of the spectrum in its transition to becoming a democracy, the "regime type" factor would have been more significant to the result. Mossadegh inherited a political institution that had been built on dictatorship, corruption, and greed, which few leaders had survived longer than a year in power. The 1953 Iran coup case study does demonstrate the association between leader characteristics, public dissent, global instability, internal conflict, and regime type as captured in the model. The presence of each factor has a direct influence on the other, albeit in varying degrees. The ensuing case studies will analyze other countries that experienced a covert action event and will attempt a connection between the factors and how they affect leader survival.

This case study analyzes the effects of five causal conditions on the fate of leaders to include leadership characteristics, public dissent, global instability, internal conflict, and regime type. Utilizing the QCA process, this case study identifies characteristics of each factor that will be used to determine the degree of membership of each case. Regarding leader characteristics, Mossadegh was a member of the minority population, he had a college degree, and he was under the age of forty years old when he assumed power. Public dissent is not as significant of a factor because Mossadegh enjoyed a tremendous amount of public support from the Iranian people. He assumed power during a tumultuous time when political corruption ran rampant and very few political leaders survived longer than twelve months in office. Looking at global instability, the threat of communism was strong and heavily influenced foreign policy behavior of the United States. Transitioning to internal conflict, the country was in a state of crisis at the time of the covert action event. Military coups occurred on a cyclic basis and the environment was ripe for a revolution. Considering regime type, the people of Iran had grown weary of the monarchy and its abuse of power as well as exploitation of the country's resources for personal gain. The tenuous relationship between the monarch, prime minister, and military leaders were influential pre-conditions to inevitable regime change.

Table 2.1.

Variable/ Membership	*Fully In*	*Mostly In*	*More In Than Out*	*More Out Than In*	*Mostly Out*	*Fully Out*
Leader Characteristics		X				
Public Dissent			X			
Global Instability	X					
Internal Conflict	X					
Regime Type					X	

Chapter Three

Guatemala, 1953–1954

By the time the 1954 Guatemala coup occurred, the country was fully immersed in political turmoil. Two major political factions were at odds with each other, which were the conservative side as represented by wealthy landowners and military officers and those who passionately fought for the left, as symbolized by the Guatemalan Communist Party. Nationalism, communism, militarism, and strong divisions within the social fabric of Guatemalan society set the stage for the 1954 coup. All of these powerful influences converged over the issue of land reform. A key stakeholder in the initiative was the United Fruit Company, an American conglomerate which, at the time, was the largest foreign business presence in the country. Shareholders of the company urged politicians in Washington, DC, to support a coup to overthrow the Arbenz Guzman presidency. So, they did, with failed results. As Cullather notes, "Ironically, by attaining its short-term goals, removing Jacobo Arbenz, PBSUCCESS thwarted the long-term objective of producing a stable, non Communist Guatemala" (1994, 91).

PBSUCCESS exploited paramilitary and psychological operations to replace a popular, democratically elected government with an unknown successor. The objective of the Dwight D. Eisenhower administration was to use covert action to replace a government with perceived communist leanings with a stable government that rejected communism and supported US economic interests. Instead, ousting Arbenz resulted in countrywide instability and laid the groundwork for the failed Bay of Pigs operation seven years later.

After removing Arbenz, the United States set out to use Guatemala as its showcase for democracy. However, other forces were far stronger than those of democratic values. This case study highlights the importance of global instability in the form of external influence asserted by the United States. It also underscores the personality profile of a confident but neurotic leader of

Guatemala, his exploitation of the agricultural resources of Guatemala, and how his actions proved to be the primary source of internal conflict in the country. The United States exercised influence and coercion over Arbenz and, during his tenure, the Guatemalan Army quickly leveraged this animosity between Arbenz and the United States to challenge his authority.

While this case illustrates the importance of the previously mentioned factors of the QCA model, the case of Guatemala does not demonstrate the significance of leadership characteristics in leader survival. Arbenz was portrayed as a socialist leader who was, in reality, leading the country under communist control. While he was not a passionate nationalist, he was easily influenced by his wife. He traveled around the Soviet Bloc countries to relocate and never settled but ensured that his children received education in Moscow, thereby ensuring that the youngest generation of the family would be indoctrinated into the family's leftist ideology. Once a promising young military officer, he never rid himself of the bitterness he felt as an outsider in Guatemalan society.

BACKGROUND

To the United States, Guatemala was on the brink of communism. To Arbenz and his followers, land reform and progressive thinking were the means to bring equality to a country that had been in turmoil since its inception. The deep chasm between the elite landowners and the workers grew deeper and wider with every leadership transition. The successful coup to overthrow Arbenz represented the vulnerability of small countries to larger superpowers. This case study documents the United States' use of economic interests to assert its influence over a foreign country to the detriment of its people. In comparison to the Iran case study, Guatemala offered natural resources in the form of agricultural products, whereas Iran had an abundant supply of oil. Both countries were exploited by foreign powers.

The decision to overthrow the Arbenz government was based on a US perception of Guatemala's favorable disposition toward communism. Many scholars have argued that the government was not communist at all. Rather, it was a progressive version of a reformist government that aimed to give land rights back to the indigenous population. Additionally, there was a perception of the CIA and the Eisenhower administration making executive decisions exclusively based on the economic interest of the United States as opposed to giving any consideration of the socio-economic disparity in Guatemala. Furthermore, the United States was not prepared for the international condemnation that it faced in the aftermath. London and Paris newspapers accused the Eisenhower administration of introducing a new form of "economic colonialism" (Cullather 2004, 85). The fallout from the Arbenz regime removal resulted in a legacy of anti-American sentiments that spread throughout Latin America. The regime change occurred quickly and with absolute support from the Eisenhower administration, the CIA, and Arbenz's successor Castillo Armas who had recently graduated as a student from the United States Army's Command and General Staff College. Naturally, Armas was viewed as a trusted friend of the United States in contrast to Arbenz who was viewed as a friend of the socialist group influenced by communist power. According to Cullather (1994), "Agency officials had only a dim idea of what had occurred in Guatemala before Jacobo Arbenz Guzman came to power in 1950" (1994, 2). Cullather adds,

> In Havana, Santiago, Mexico City, Buenos Aires, and Rio de Janeiro, large crowds gathered to burn the stars and stripes and effigies of Eisenhower and Dulles. Even before the afterglow of the White House briefing wore off, the Eisenhower administration had reason to question whether PBSUCCESS had delivered an undiluted victory. Agency and State Department officials were shocked at the ferocity of international protest after the fall of Arbenz. (2004, 85)

The Eisenhower administration ultimately recognized the reality of the situation and saw events not just in the Guatemalan context but as a part of a global pattern of communist activity. Instead of witnessing the blossoming of a stable government which would, in turn, lead to a favorable economic climate, many began to see Guatemala as it truly was a politically unstable country ravaged by decades of internal conflict; a country deeply divided by cultural and religious beliefs; a government that provided inadequate living conditions for a majority of its population; rampant poverty and illiteracy; and a prime candidate for US foreign aid. As Cullather asserts, "Resentment even found artistic expression in the work of Mexican muralist Diego Rivera, who depicted in fresco who painted [US Ambassador] Puerifoy and the Dulles brothers passing money to Castillo Armas . . . over the bodies of Guatemalan children" (2004, 86).

In Booth's (2014, 175) book on Guatemala's history, he notes that "Arbenz sought to deepen social reforms despite growing conservative and US opposition. In 1951 he legalized the Communist Party, called the Guatemalan Labor Party (Partido Guatemalteco del Trabajo, PGT)." This organization began actively pushing for agrarian reform under the guidance of Arbenz and his predecessor. According to Booth (2014, 175), "The 1952 Agrarian Reform Law began the confiscation and redistribution of farmland to 100,000 peasants." The new strategy had seismic effects on the local populace and beyond.

For the first time in the country's history, economic power shifted from the one percent of elite landowners (most of whom were foreign) to workers and peasants from the indigenous population. The new regulations led to the nationalization of land belonging to the United Fruit Company (UFCO).

At the time, the UFCO was described as "nearly 300,000 acres of banana and sugar cultivations spread over six countries and one British colony" (Martin 2013, 238). Founded by Texas entrepreneur Minor Cooper Keith in 1899, UFCO created a vast caste-based wild frontier in which members of the top hierarchy included a nucleus of white, Anglo-Saxon men who served as managers and supervisors. Lower-class laborers were mostly Hispanic and West Indian.

Martin (2013, 238) noted, "Throughout this period, there were daunting medical obstacles to retaining a stable population of what company doctors came to term 'temperate zone' employees." He added, "Reminiscent of North American frontiers, United Fruit's raw edges were violent places and, true to the frontier metaphor, disorder emanated primarily from people of color" (2013, 322). The term "banana cowboys" came to describe the men who worked the Guatemalan banana trade. They represented the banana frontier and "its unrestrained male sociability, its transformative possibilities and the racialized labor discipline that kept the business running" (Martin 2013, 328).

Considered an economic paradox, Guatemala was a country blessed with abundant agricultural resources. The nationalization of the UFCO raised red flags for the Eisenhower administration that had close ties with the conglomerate. Nationalization, inklings of communist affiliation, and Guatemala's coincidental purchase of small arms from Czechoslovakia indicated to the US that the country was becoming increasingly communist. The memory of the Iran coup over the threat of communism still lingered in the collective memory of many. In response to Arbenz's agricultural reform strategy and nationalization of the UFCO, the United States implemented a plan to destabilize the Arbenz administration through the use of economic sanctions, diplomatic influence, propaganda, and psychological operations. As Trefzger (2002) notes, "In June 1954 the tiny CIA-supported National Liberation Army led by Colonel Carlos Castillo Armas, invaded Guatemala. The armed forces refused to defend the government, and President Arbenz had to resign" (Booth 2014, 175). Guatemala's destiny of becoming a democratic nation had been cut short.

LEADER CHARACTERISTICS

Born in 1913 in Guatemala, Arbenz was the son of a Swiss pharmacist who immigrated to Guatemala and eventually married a local Guatemalan woman.

His father's pharmaceutical business failed as a result of his morphine addiction and the Great Depression, and he committed suicide in 1934. Described as "tall, good-looking and motivated by an intense desire to make up for his father's failure," Arbenz was a hard-working student who devoted many long hours to his studies. He graduated at the top of his class at the military academy but quickly came to realize that the traditional military structure put into place by the current regime offered little, if any, upward mobility for a person from low economic and social class. As Immerman suggests, "Frustrated, Arbenz came to hate the system that denied success to an individual simply because he was from the wrong class. He gravitated toward the revolutionary movement, and, after Ubico forced him into exile at the age of thirty, he became one of the leaders of the opposition" (1982, 1212). With his engaging personality and vibrant voice, he soon became a favorite among the Guatemalan people. He had all of the ingredients for success: intellect, passion for national reform issues, skilled rhetoric, and a beautiful wife who had political connections of her own as a member of one of El Salvador's most wealthy families. According to Immerman (1982, 1212), "Maria's father was associated with the government's 1931 massacre of nearly twenty thousand Indians and members of the indigenous population. Horrified by what she had witnessed, the young Maria adopted an almost parental concern for the Maya peasants." Together, they formed the perfect image of the country's young and attractive "First Family" of Guatemala. However, the perfect image would be short-lived, as the United States grew concerned over Arbenz's favorable leanings toward socialist theory. The threat became so extreme that discussions over assassinations and leader removals were presented within the inner circle of Washington leaders.

From *CIA and Guatemala Assassination Proposals, 1952–1954, CIA History Staff Analysis by Gerald K. Haines, June 1995*, a declassified memorandum from the CIA states,

> In addition to monitoring events in Guatemala, the Agency continued to try to influence developments and to float ideas for disposing of key figures in the government. It proposed to not only focus on sabotage, defection, penetration, and propaganda efforts with regard to Guatemala, but to also eliminate and assassinate key Guatemalan military leaders. (1995)

Further, the memorandum also adds, "The CIA plan, to draw up the Western Hemisphere Division, combined psychological warfare, economic, diplomatic, and paramilitary actions against Guatemala. Named PBSUCCESS, and coordinated with the State Department, the plan's objectives was to remove covertly, and without bloodshed if possible, the menace of the present communist-controlled government of Guatemala" (1995, 4).

In regard to training, the memorandum notes,

> Although assassination was not mentioned specifically in the overall plan, the Chief . . . requested a special paper on the liquidation of personnel on 5 January 1954. A cable [delivered] the following day requested 20 silencers, for .22 caliber rifles. The chief also discussed the training plan that would require the training of two assassins, also known as "assassin specialists." The memorandum also revealed that, "Intense psychological warfare techniques were utilized as well. The propaganda campaign was focused on the Arbenz administration and included sending mourning cards to top communist leaders, letters indicating the impending use of nerve agents, wooden coffins, hangman's nooses and phony bomb were delivered to leaders' homes." A CIA station chief added, "The elimination of those in high positions of the government would bring about its collapse. He added, even a smaller number, around 20, would be sufficient." (Haines 1995)

On 27 June 1954, Arbenz, in a bitter and anti-American speech, resigned his presidency and left for Mexico where he was granted asylum. He was only the second president in Guatemalan history to be democratically elected to office. His predecessor, Juan Arévalo was elected through a democratic process and won with 85 percent of the popular vote. Aybar de Soto (1978) writes, "After coming into power, Arbenz continued most of the reform programs initiated by his predecessor. Most of the reform programs of Arbenz proceeded within the broad framework of democracy. Many scholars evaluated his reform programs as substantially democratic. For approximately ten years, the fight for organizing a labor union and reforming a program of land distribution continued with a desire for establishing a foundation of mass support that would provide institutional bases of democracy" (de Soto 1978). Based on the five-factor model of personality and transformational leadership, this section will assess Arbenz's leader characteristics in the areas of neuroticism, extraversion, openness to experience, agreeableness, and conscientiousness. Information captured from a declassified memorandum released by the CIA in 2003 offers an intimate biographical profile of Arbenz during the period leading up to and following the coup.

NEUROTICISM

As noted in the previous case study, the trait of neuroticism is typically associated with individuals who lack self-confidence and self-esteem. The following excerpts allude to Arbenz's unsteady personality and lack of personal discipline and self-control. It appeared that he progressively lost confidence with the passing of time.

In the CIA character profile, compliments of the CIA Historical Review Program, the details include (https://www.cia.gov/library/readingroom/docs/DOC_0000919960.pdf):

28 June 1954. Arbenz, wife and two children took refuge in the Mexican Embassy in Guatemala. This was followed by numerous rumors of Arbenz being ill, and many reports of his moodiness and drunkenness within the Embassy.

10 July 1954. According to diplomatic sources, Arbenz attempted to commit suicide in the Mexican Embassy in Guatemala City on 10 July.

19 October 1955. Antonio Vilanova, Arbenz's brother-in-law, said during a recent trip to Mexico that he doubts that Arbenz's wife will permit him to return to Mexico, since she was responsible for their going to Europe in order to have him recover from his mental depression. Vilanova said that Arbenz had tried to commit suicide a number of times, that he has become a heavy drinker, going on periodic, prolonged bouts of intoxication.

November–December 1956. Arbenz continues his close relationship with José Manuel Fortuny. Arbenz gets drunk and his conduct is poor. The difficult relationship with his wife, her infidelities, the possibility of an increase in these problems, and his being alone in Paris—a situation he calls a hopeless life—cause him to drink to excess, alone or in company. Arbenz has always been a neurotic. His desperation causes him to be locked in his room for days on end at the Hotel Vermont, Paris, having food sent to his room, not speaking to anyone, with the windows closed and the lights turned off day and night. He spends his time in absolute depression, violent irritation, and crying spells. Physically, Arbenz is exhausted and looks old. His character makes him more impulsive and violent. He seems to be a man without strength, without the desire to live; or at least one who wants to live peacefully without fighting. Arbenz revealed a certain reluctance to continue an intensive political life.

24 December 1956. Arbenz was arrested for drunkenness by the French police on Christmas Eve.

31 March 1957. Arbenz's wife wrote him from El Salvador expressing concern that she has had no information from her daughters in two months. There is some cross play in the letter indicating that they are accusing each other of lack of tenderness in their letters. She states that they are no longer in the Romeo and Juliet phase. She comments that his concern over the tone of her letters may be due to the state of his nerves.

EXTRAVERSION

Extraversion is related to "leader emergence in groups" (Watson and Clark 1997, 767). Based on the personality profile, it does not appear that Arbenz was an extroverted leader. In the months following the coup, he became increasingly more introverted and depressed.

> **19 September 1954**. Arbenz has been very depressed. This fact is confirmed by the very few who have seen him (his assistants, Guillermo Toriello and Augusto Charnaud MacDonald). Few of the people who sought asylum in the Mexican Embassy have been permitted to visit Arbenz, but he has not been ill as stated by the press. (https://www.cia.gov/library/readingroom/docs/DOC_0000919960 .pdf)

OPENNESS TO EXPERIENCE

Scholars describe the trait of openness to experience as being associated with creativity and open thinking. Arbenz demonstrated this trait through his experience of traveling around the Soviet Bloc countries in order to find a home for him and his family.

> **19 September 1954**. Jacobo Arbenz, ex-president of Guatemala, and his wife have decided to go to Switzerland as soon as possible once they have obtained their safe conduct permits to leave Guatemala. Arbenz's parents were born in Switzerland. Mrs. Arbenz has been studying German in the Mexican Embassy in Guatemala.

> **20 April 1956**. Maria de Arbenz in a letter from Moscow dated 20 April 1956 said that she and her husband would leave for China about the middle of May 1956. She said that their future plans would be known in about three months. At this time consideration was being given to having Arbenz travel to India, Indonesia and North Africa for the purpose of building himself up politically and serving the Soviets by inspiring national liberation movements. Arbenz and his wife might live outside the Soviet orbit. Arbenz prefers Egypt, but his wife prefers Italy. Arbenz dislikes France. (https://www.cia.gov/library/readingroom/docs/DOC_0000919960.pdf)

AGREEABLENESS

Charismatic leaders have been described as generous and concerned for others. There are indications in the personality profile that indicate Arbenz's tendency to be compassionate towards others. There are even more periods when he demonstrates a high level of disagreeableness.

27 October 1954. The weekly magazine *Siempre* carried an interview with Arbenz in its 27 October 1954 issue in which Arbenz accused the Guatemalan Army of betraying him. Comment: This and similar anti-Army themes should be stressed in internal propaganda in Guatemala.

22 August–24 October 1955. It is rumored among Guatemalan exiles that Arbenz received a million dollars from the USSR for use in Mexico and Central America for propaganda. The reported trip of Arbenz into the Soviet Bloc has raised the hopes of the Guatemalan Communist exiles. Arbenz is believed to be in touch with Communists in Mexico, Cuba, and Central America.

25 November 1955. On 25 November 1955 Arbenz stated to Western journalists in Prague that he did not plan to return to Guatemala unless the government imposed by the American imperialists is replaced by a new, really democratic government. (https://www.cia.gov/library/readingroom/docs/DOC_0000919960.pdf)

CONSCIENTIOUSNESS

Conscientiousness is described as an awareness of importance to the people. Unlike Iran's Mossadegh who knew that he was a well-liked, popular leader, Arbenz had support from the local population but appeared to have more interest in pursuing communist doctrine than understanding the needs of the Guatemalan people. There are moments that indicate his sensitivity to protecting the safety of fellow exiles. Following the coup, he declined to state that he was concerned his comments would have a negative impact on the status of exiles in Mexico.

14 October 1954. Arbenz replied he did not think it was an opportune moment because of the promise Mexico had requested from him to abstain from any political action which might cause a strain in relations between Mexico and Guatemala . . . any statements from him to the press at this time would increase the campaign against Guatemala exiles and would give the participants in this campaign new reason for requesting the extradition of these exiles. (https://www.cia.gov/library/readingroom/docs/DOC_0000919960.pdf)

This indicates some sense of compassion towards his fellow exiles.

25 November 1954. The exile group in Mexico is not happy with Arbenz's planned trip to Switzerland because he has not divided the money taken

> from the Guatemalan Treasury prior to his resignation. The money had been deposited in Swiss banks. (https://www.cia.gov/library/readingroom/docs/DOC_0000919960.pdf)

This excerpt indicates that he was only in it for himself and was fully aware of his actions. His lukewarm reception back into the country demonstrates a strong lack of support from his fellow citizens.

> **23 June 1955**. Guatemalan exiles in Mexico expect Arbenz to return from Europe in the immediate future. Many members of the Partido de Accion Revolucionaria Unificado dread his return because they fear it may create a greater division between the Guatemala exile groups. They fear, in particular, his favoritism towards the communists. (https://www.cia.gov/library/readingroom/docs/DOC_0000919960.pdf)

PUBLIC DISSENT

Arbenz came into office on a wave of public support. He was a popular nationalist leader and his removal resulted in mass protests not just within but also outside the country. As Booth notes,

> Reformists under Guatemala's democratic governments of 1944 to 1954 suffered badly after the 1954 coup. The military's demobilization campaign decimated the ranks of reformist politicians, unionists, and indigenous people who had supported the Arévalo and Arbenz governments. Marxist guerrilla opposition to the regime first appeared in 1962, but was set back greatly by heavy general repression and an intense counterinsurgency campaign in the late 1960s. Popular mobilization rekindled during the 1970s as real wages and income distribution worsened. (2014, 178)

Booth adds, "During the post-1954 counterrevolution, Guatemalan rulers intensely repressed union activists, students, peasant groups, indigenous peoples, opposition parties, and other dissidents" (2004, 179). The reform strategy may have gained traction through the popular support of President Arbenz. However, the initiative to improve healthcare and raise literacy rates among the indigenous population was only a Band-Aid that concealed a deeper wound: gross misdistribution of land between the small percentage of elites and the large population of Indians. The covert action event that ousted Arbenz would thwart any hope for democracy, independence, and equality for the people of Guatemala.

GLOBAL INSTABILITY

International superpowers were riding a crest of so-called success with the removal from power of Mossadegh in Iran, all in the name of thwarting communism. Overly confident and riding the wave of success, the members of the Eisenhower administration were ready to expand influence on another country with the potential for economic gain.

As Sewell contends,

> In the 1950s, the Cold War was often presented as a battle of ideas, yet CIA agents on the ground didn't see it that way. They insisted on a strategy intended to inspire fear more than virtue. During the period when Dwight D. Eisenhower was president and Allen Dulles was the Director of Central Intelligence (DCI), disparate sources have argued, the CIA enjoyed a position of unparalleled prominence in US foreign policy. With Eisenhower's proclivity toward psychological warfare and covert operations increasingly to the fore, the CIA was an essential part of his armory in the ongoing struggle against the Soviet Union. (2011, 270)

"The Soviet Union," John Foster Dulles, secretary of state, quickly concurred, "had cleverly infiltrated mass political movements in Latin America" (Sewell 2011, 283). Sewell adds, "Allen Dulles [director of the Central Intelligence Agency and brother of John Foster Dulles] may have been a key protagonist in building the Cold War CIA, but his tenure also oversaw a series of disastrous and illegal interventions. Dulles's Agency broke international law, subverted democratically-elected governments and supported brutal dictatorships as long as they were pro-American—all under a banner of legitimization provided by the Cold War" (Sewell 2011, 270).

INTERNAL CONFLICT

Deep divisions between the small population of elite landowners (mostly who were foreigners) and uneducated indigenous populations created conflict within the country.

Fraser writes,

> When former Guatemalan Army officer Jacobo Arbenz ascended to the country's highest office in 1950, it marked the first time in Guatemala's 130-year history as a nation that the reins of governance had passed from one democratically elected leader to another. Arbenz rode to power on the crest of a wave originating from the national revolution of October 1944. Arbenz, an army captain who had resigned only months earlier, and his colleague Francisco Arana, a colonel in the Guatemalan Army's most renowned unit, the Guardia de Honor,

> launched a bloody revolt against their commanders. Under intense pressure, Ponce resigned. Arbenz, along with Arana and prominent anti-Ponce businessman Jorge Toriello, established a temporary government and set out a timetable for elections. It would be the nation's first, albeit rudimentary, experiment with democracy. (2005, 487)

Guatemala's experiment with democracy was led by a socialist-minded, charismatic leader by the name of Juan José Arévalo, a professor of agricultural science who became the first civilian to lead the nation—an outlier in a country where power fell mostly into the hands of dictators and military officers (oftentimes one and the same). It was during this time that Jacobo Arbenz was appointed defense minister in Arévalo's new administration.

Since the coup, the country of Guatemala has experienced insurgency, political revolt, civil war, genocide, and currently serves as a battleground for criminal elements and violent gangs. Brands writes, "During the Cold War, longstanding social, political, and economic strains mixed with the ideological polarization produced by superpower rivalry to generate torrential bloodshed in Guatemala, El Salvador, Peru, Colombia, and elsewhere. The worst of these conflicts wrecked economic capacity and government institutions, and set Latin American countries back by decades" (2011, 230).

REGIME TYPE

Guatemala's regime type can be described as dictatorship, authoritarian regime on the brink of democracy, culminating in a democratic election that led to a US-orchestrated overthrow.

Brands adds,

> The end of the Cold War brought down the curtain on many of these insurgencies, but it did not bring an end to internal violence and upheaval in Latin America. Over the last two decades, an array of criminal groups—notably international drug traffickers, violent youth gangs, and organized crime syndicates—have emerged as the chief purveyors of disorder. Across the region, these groups have exploited weak institutions, official corruption, porous borders, inequality and social alienation, and the easy availability of small arms to expand their operations dramatically. They have carved out nodes in a variety of illicit activities—drug smuggling, human trafficking, arms dealing, kidnapping, robbery, extortion, money laundering, and others—and made insecurity a fact of life. Latin America regularly competes for the title of most violent region in the world (in terms of homicides), and its youth murder rate was more than twice as high as that of any other region in 2008. (2011, 230)

The roots of contemporary violence in Guatemala can be found in the culture of ruthless dictatorships, greedy landowners, abuse of military and police authority, and influences from external powers. The clandestine intelligence and security apparatuses that were created during the Cold War have never been completely dismantled and, like weapons left behind by superpower countries, these organizations continue to evolve in the new world.

As Fraser (2005, 487) notes, "The American attempt to depose the Guatemalan regime would certainly have failed had the Guatemalan military not abandoned Jacobo Arbenz. Aspirations for peace and freedom ultimately failed in Guatemala, now to a greater degree than before the architecture of that story can be revealed, as seen from inside the American government. It is the architecture of a broken dream."

The case of Guatemala demonstrates that a leader's characteristics and personality traits are necessary for survival. Had Arbenz been more confident of a leader, more savvy about foreign relations, and had made an effort to negotiate with the United States, he may have fared better in the end. Instead, he left his country to find residence in communist-friendly locales—a message that was viewed with much suspicion by the CIA and the Eisenhower administration.

The 1954 Guatemala coup case study shares similar themes with the previous case study on Iran's 1953 coup. Both countries were viewed under suspicious auspices by the United States for communist leanings, and both countries were blessed with bountiful natural resources. In both cases, external powers asserted economic and political control over the country's leaders, resulting in their untimely removal from office.

It is interesting to note that the Guatemala case study highlights global instability and internal conflict as the strongest factors in influencing leader survival. Arbenz enjoyed popular support and sought to improve living conditions for the indigenous population of Guatemala. While his leadership characteristics were not necessarily strong, per se, it is assessed that he was easily influenced by external factors. For example, his wife, Maria, was a critical factor for his interest and eventual obsession with communist theory and doctrine. Together, they shared a passion for social reform and a new brand of democracy—at least in their own minds. Arbenz's abrupt removal from office triggered internal conflict, and external forces shaped his direction towards communist associations.

The US role in the Guatemalan coup was viewed as a victory by the United States. As with previous cases, the Guatemala coup case study also analyzed the effects of five causal conditions on the fate of leaders. Considering leader characteristics, Arbenz was a member of the middle-class and son to a Swiss expatriate and Guatemalan mother, he was a graduate of the country's most elite military institution, and he was under the age of forty years old when he assumed power. Public dissent was divided with more than half of the population supporting Arbenz and several political groups wanting to oust him from power. He assumed power during a tumultuous time when land reform was at its most contentious with a small one percent owning the majority of arable land. Global instability was an influential factor as the world was greatly divided between democracy and communism, and national leaders who did not fall in line with democratic values were often viewed with suspicion by the United States. Internal conflict played a large role as the country was in a state of crisis at the time of the covert action event. There had been a revolution over land reform and rights for the minority population. One of the most successful aspects of this case study was the ample information provided by the CIA on Arbenz's personal life. It seemed that every detail about his life was documented and recorded for over a decade until his death.

Table 3.1.

Variable/ Membership	*Fully In*	*Mostly In*	*More In Than Out*	*More Out Than In*	*Mostly Out*	*Fully Out*
Leader Characteristics				X		
Public Dissent			X			
Global Instability	X					
Internal Conflict	X					
Regime Type				X		

Chapter Four

Indonesia, 1955–1958

In the 1950s, Indonesia was in the crosshairs of the United States and the Soviet Union; the first through democracy and freedom and the latter through communism as both countries competed for political and economic influence on the global stage. The United States' foreign policy at the time, both globally and in the Asian region, was driven by containment of the communist threat. Indonesia, in its decision to maintain neutrality was viewed with much suspicion and, to add to this suspicion, the newly independent nation of Indonesia was experiencing a surge of membership in its Indonesian Communist Party (the PKI). Moreover, members of the PKI were gaining coveted positions within the political institutions led by Indonesian President Sukarno. The United States responded to this influx of communist influence with the implementation of its two-track foreign policy: the first option of persuading Indonesia to pull away from neutrality by working through official diplomatic channels at the embassy in Jakarta, and the second option by covert action through insurgents located throughout Indonesia's outer islands.

It was also during this time that the CIA was riding the crest of recent covert action successes in Iran and Guatemala. With President Dwight D. Eisenhower's approval, the CIA was preparing to initiate another operation in Indonesia under the code name Operation HAIK. Unlike Iran's democratically elected Prime Minister Mohammad Mossadegh who had his country perched on the edge and leaning more toward democracy and Guatemala's democratically elected President Jacobo Arbenz who implemented land reform programs with a socialist flavor, Indonesia was not close to becoming a democracy and nor did its leader, Sukarno, have plans to make it a democracy. Instead, he saw a path of what he called "guided democracy," a cross between democracy and socialism, and he would be the plan's guide for the people of Indonesia.

This case study highlights the importance of leader characteristics as well as the power of internal conflict, stemming from competing political factions, and the persuasive and strong-arm influence of a country's military and security apparatus. As seen in the previous two case studies, global instability was primarily created by international superpowers, such as the United States and Soviet Union, and the divisive line drawn between the democratic and communist ideologies. Based on the study's model, the case of Indonesia incorporates all the facets of leader characteristics, public dissent, global instability, internal conflict, and regime type. The outcome of the 1958 Indonesian failed coup opened the Pandora's box for a myriad of socio-cultural and security issues for the country for many decades to follow. The poor execution of coups, more often than not, provides an impetus for a country's transition to internal turmoil, dissent, radical terrorism, and human rights issues as illustrated in Indonesia post the 1958 failed coup attempt.

BACKGROUND

As the world's most populous Muslim majority country and the largest archipelago in Asia, Indonesia is well-known for its beautiful scenery, exotic culture, and brutal history of religion-sparked violence, political corruption, and radical Islamic terrorism. The history of Indonesia is complex, especially within the context of covert action. As a beautiful and relatively peaceful archipelago in Asia, the country was a mystery to most foreigners. Its mysticism and mix of cultures made it a land of intrigue and its natural resources were in abundant supply. It seemed like the perfect location for European colonialists to settle, and that is exactly what they did. It also became a haven for Islam, a religion that was introduced to Indonesia in the eleventh century, which quickly spread throughout the outer islands and, eventually on the main island. Ethnic divisions and political factions obstructed Indonesia from becoming a unified country. More importantly, the divisions and cultural differences made it fertile ground for exploitation by external forces.

As Frederick and Worden note in *Indonesia: A Country Study*, "Although Indonesia is extremely diverse ethnically (more than 300 distinct ethnic groups are recognized), most Indonesians are linguistically, and culturally, part of a larger Indo-Malaysian world encompassing present-day Malaysia, Brunei, the Philippines, and other parts of insular and mainland Asia" (1993, 1). The two most significant developments in Indonesia's history under the Sukarno government were its quest to gain land in West New Guinea and development and growth of Indonesia's military force. As it turns out, the latter became the most successful and powerful effort out of the two.

The battle of West New Guinea was a common platform with which a majority of Indonesians could identify. Having won independence from the Dutch, the people of Indonesia were stalwart in their claim over land they deemed was rightfully theirs.

Indonesian culture is a complex blend of Javanese tradition, mysticism, and Islamic influence, further complicated by ethnic and cultural divisions between the members of the island and those who lived on the smaller islands. The influence of Islam was also a powerful force that created deep religious divisions. As Frederick and Worden note (1993, 4), "Because of the antiquity of Java's civilizations and the relative isolation of some of its most powerful kingdoms, the process of Islamization there was both complex and protracted." The presence of Dutch colonialists also created another opportunity to blend Eastern and Western ways. A select few of Indonesia's elite class were afforded the opportunity to send their sons to Dutch primary schools. This opportunity was only offered to those who enjoyed social position and influence. Many of the Indonesian youth chosen to attend Dutch schools grew up to be men who were influential in local and national politics. They recalled the days when they experienced discrimination from their Dutch peers, and they used that collective memory to shape their views of politics and society as a whole. Many, like Sukarno, had a deep dislike for the Dutch and believed they represented a threat to Indonesia's independence from colonialization. Sukarno believed that Islam, specifically Saraket Islam, which was introduced to the island of Indonesia during the Dutch colonialism years, was a religion that represented the downtrodden and, thus, he associated it with the nationalist cause. Sukarno had a deep passion for unifying the influences swirling around him. He believed that he could use elements of Marxism and Islam to bring together the people of Indonesia, despite the varied culture, ethnic and religious divisions, and general mistrust between the city and village populations and those who inhabited the outer lying islands.

A key theme present in this research is Sukarno's passion for theatrics and using his charisma and rhetoric to influence the masses. He believed he could use his oratory gifts to synthesize ideas and mobilize the masses as, ultimately, he found his words needed powerful actions behind them. After completing his engineering degree, Sukarno was on his way to becoming a political leader. He had the innate ability to mesmerize any audience, and many Indonesians viewed him as a visionary leader. Sukarno's downfall was his inability to deal with the reality of leading a country, navigating through the turbulent waters of cultural/religious divisions, and skillfully representing himself as a competent leader in the international arena. While he was idolized by many Indonesians as a visionary leader and nationalist, his critics knew his weakness was in vanity. As Frederick and Worden (1993, 7) noted, Indonesia's ailing economy grew worse as Sukarno built expensive public monuments of himself and government buildings to celebrate his position and status.

LEADER CHARACTERISTICS

Sukarno was born on 6 June 1901 in Surabaya, Java, and Dutch East Indies. He held the honor of being Indonesia's first president and led the people during the country's first independence movement. His vision of a guided democracy and his determination to maintain neutrality in the midst of Cold War tension sealed his legacy as a charismatic and confident leader.

As the son of a poor Javanese schoolteacher and his Balinese wife, Sukarno was known by his peers as one who had leader qualities despite his pattern of ill health. He was a loyal and devoted follower of the art of Wayang shadow puppetry, which told the story of Indonesia with nothing but paper and bedsheets. Sukarno became the protégé of a local religious leader who funded his schooling and eventually married him off to his sixteen-year-old daughter. As a student, he had a natural ease with languages, including Javanese, Sundanese, Balinese, modern Indonesian, Arabic, Dutch, German, French, English, and eventually Japanese. According to his biographers, he was blessed with a commanding presence, a radiant personality, a photographic memory, and self-confidence.

During his tenure as president, he was well-known for his elaborate banquets, nightly soirees, music, and Wayang theatrical performances. Meanwhile, the Indonesian economy continued its perilous descent. While Sukarno had a strong vision of unity for the country, cultural, moral, religious, political, and economic differences created a steady state of disunity. Separatist movements also created internal instability. As the population continued to increase, food production began to decrease as a result of the destruction of plantations during the war. A majority of the population was illiterate, and the lack of skilled workers also wreaked havoc on a war-weary economy. His critics claimed that Sukarno's guided democracy was touted as the antithesis of democratic values, but the reality was that it leaned more towards the communist ideology.

Sukarno has been described as everything from a nationalist leader and hero of the people to disastrous president who dismantled the framework for democracy in the 1950s, the source of confrontational foreign policy relationships with Malaysia in 1963, and the leader responsible for leading Indonesia down an economic spiral from which it has yet to recover. On the other hand, many people remember Sukarno as a gifted leader who had the ability to relate to and communicate with people from all social classes.

According to Legge,

> He was a highly individual leader with his own distinctive and mercurial style. His vanity, his charm, his political resourcefulness and his unpredictability were a source both of admiration and exasperation to observers. Among his countrymen he inspired devotion or hostility, but never indifference. Even his most consistent critics were fascinated by him in spite of themselves. To some he was a dedicated leader—a real maker of Indonesia and the sustainer and preserver of the nation after independence. To others he was a disaster, the waster of the nation's resources in policies of domestic extravagance and foreign adventure, a man seduced by power and by the material pleasure of presidency, the corrupter of his people and himself. (1972, 52)

As Judge and Bono note in their work on transformational leadership, "Idealized influence can be defined as serving as a charismatic role model to followers. Inspirational motivation involves articulation of a clear, appealing, and inspiring vision to followers" (2000, 751). Sukarno certainly had the ability to influence people from all cross-sections of Indonesian society, ranging from the village farmer to the military and members of the elite political circle. He appeared to be committed to unifying the country and leading it toward independence. As Eden (1992, 271) noted, "Setting high performance standards and then convincing followers that attainment of these standards is possible lies in the heart of transformational leadership." House and Howell (1992, 87) added, "Theoretically, charismatic leaders need to have a very high degree of self-confidence and moral conviction because their mission is usually unconventional and likely to be resisted by those who have a stake in preserving the status quo."

NEUROTICISM

Sukarno's relationship with the people of Indonesia was shaped by his belief in mysticism and his passion for theater. He created an image of himself that was, in essence, mystical. His ability to put on a new persona or take on a new role—much like an actor in a theater production—allowed him to interact with the population in a myriad of ways. The people responded to his chameleon-like personality and their perception of him as a lauded leader was firmly ensconced in the public's mind.

A memorandum from Mr. Walter S. Rogers, Institute of World Affairs, dated November 20, 1954:

> When Sukarno speaks, the eyes and minds of the audience seem to be riveted on his commanding figure and each turn of phrase in his speech. Critics claim that Sukarno has an unequalled knowledge of the habits, hopes, and prejudices of each group he addresses, whether it be Javanese, Sudanese, or Dayak, hence his great success in formulating and manipulating nationalistic opinion. (1964, 1)

Legge noted in the book *Sukarno: A Political Biography*,

> Sukarno's most noticeable skills were those of exhortation and persuasion. As an orator he could hold an audience in the hollow of his hand. By his words he sought to bridge differences, to bring together diverse elements within the nation and to give them a sense of common identity. He was sensitive to subtle changes in the political climate and was uniquely able to catch and formulate the frustration of the masses. (1972)

Sukarno has been described by scholars as a complex person who viewed himself as a revolutionary leader. His exceptional ability to build rapport with and relate to individuals through personal associations made him one of the most influential leaders of the twentieth century.

Legge continued,

> The striking characteristics of Sukarno . . . were his tremendous self-assurance. [He was] always at the center of events, uniformed and neat in appearance, brisk in his movements, sparkling in his assumption of his central role, he seemed to dominate every situation. He read widely and identified himself with revolutionary leaders of the past and displaying a strongly romantic element in his sense of history. (1972)

OPENNESS TO EXPERIENCE

Conger and Kanungo note, "Charismatic leaders are not group facilitators like consensual leaders, but they are active innovators . . . their behaviors must be novel, unconventional, and out of the ordinary" (1987, 643). Sukarno embraced openness to new experience when he moved from the small village where he grew up to the cosmopolitan city of Surabaya. As Legge (1972) notes, "The move from a small country town to a large commercial city must have been disturbing as well as exciting for Sukarno, and his high hopes were no doubt tinged with anxiety as he embarked on his new experience." Imagine a city that served as a melting pot to Dutch expatriates, a thriving Chinese population, and many more Indonesians converging into the city seeking work in the burgeoning labor industry. It was in this environment that Sukarno witnessed the convergence of ideas and thoughts that inevitably shaped his view on politics and the world in general. Legge (1972) adds, "He became an omnivorous reader and, shut away in his room, he introduced himself with growing excitement to a wide range of political thought—to Jeffersonian democracy, to the Fabianisms, to Marxism. All of these contributed in different ways to shaping his thought."

The influence of city life, his mentoring from his adoptive housefather, and the racial discrimination he experienced through his childhood shaped his view on life. His first marriage was to the daughter of his mentor, Dr. Tjipto Mangoenkoesoemo, and was more of a sign of mutual respect to Mangoenkoesoemo than it was for love. Whereas Mossadegh was fueled by nationalist pride for Iran and Arbenz was driven by a passion to help the indigenous population of Guatemala, Sukarno, by contrast, seemed to have been driven by his own vanity and a deep hatred of the Dutch. "He saw the necessity for unity and began to feel his way towards the idea of an all-embracing mass organization as a means of developing a power which could challenge that of the colonial regime" (Legge 1972). To Sukarno, Indonesia was a "soul arising from a common history and from a desire and urge to live as one. Moved partly by passion and partly by consideration of political utility, Sukarno was quick to see this as a potentially unifying idea which could be used to bring together opposing factions" (Legge 1972). Hanna adds an excerpt from one of Sukarno's passionate speeches:

> The burning desire to set my people free was beyond mere personal ambition. I was consumed with it. It permeated my whole being. It filled my nostrils; it coursed through my veins. It's what a man serves his whole lifetime for. It was more than a duty, more than a calling. For me it was a—religion. (1964)

Legge (1972) comments, "Even before he made his final choice of a political career, his romanticism had made him see himself as a great leader, moving the masses, providing a national ideology and in the forefront of the struggle against the oppressor."

AGREEABLENESS

Judge and Bono (2000) describe "agreeableness" as a leader's ability to show concern towards his or her followers. Bass (1985) adds, "Transformational leaders give special attention to neglected group members, treat each subordinate as an individual, and express appreciation for a job well done."

As Mr. Walter Rogers noted after witnessing one of Sukarno's public speeches,

> Sukarno not only understands his audience, he quite obviously identifies himself with it. Speaking to farmers, he is the Javanese aristocrat and understanding father of the people. With political leaders, he is the revolutionary comrade. With university students, he forgets his role as papa and joshes or talks heart-to-heart as an equal. With Moslems, he is wholeheartedly Moslem. (1954)

Perhaps it was Sukarno's passion for Wayang theater as a child and young adult that taught him how to play different roles? Nevertheless, he was able to seamlessly transition into a character that suited the needs of the situation at the time. In his mind, his greatest role was that of revolutionary leader of Indonesia.

CONSCIENTIOUSNESS

Conscientiousness is based on a strong sense of self-determination. Moreover, achievement and self-discipline are key components of conscientiousness. Sukarno enjoyed many achievements as a student in the Dutch primary school. He had an innate ability to perform for a myriad of audiences, depending on the context and situation.

Sukarno enjoyed his reputation as a dominant leader among his peers. As a youth, he was known to lead games, organize activities, and place himself in the center of all the activity. He remarked that he asserted himself because he did not believe anyone had the confidence to usurp his authority. He appeared destined to become a national leader even at a young age. While he felt the sharp pain of discrimination from the Dutch schoolchildren, he overcame this by focusing on his sexual prowess, especially when it came to chasing and seducing the Dutch girls in his age group. While discovered by his father in a compromising position, his father merely replied that he hoped Sukarno was learning the Dutch language more efficiently as a result of his dalliances. The evidence suggests that all of these experiences culminated into a powerful force that eventually thrust him into the political realm—a world that he had prepared for his entire life. He had a reputation for being self-assured, confident, and intelligent. That foundation, combined with this experience in the urban environment as a teenager, created him into the role of a national leader.

PUBLIC DISSENT

Considered a charismatic president, Sukarno was well-known among his supporters as a leader who was anti-Western colonialism and was driven by a passion for unity and independence. Easter adds, "Sukarno openly aligned himself with the communist bloc in this struggle, proclaiming support for the North Vietnamese in the Vietnam War, establishing close ties with the People's Republic of China and angrily pulling Indonesia out of the United Nations in January 1965" (2005, 55). At the same time, Sukarno implemented a policy based on confrontation in which he publicly denounced Malaysia as a creation of British colonialism. Easter comments, "To leaders in Washington, London and Canberra, Sukarno appeared to be mounting a comprehensive challenge to Western interests in South-East Asia" (2005, 55). In essence, he was becoming the country's revolutionary leader. He used the influences of Wayang theater, Islam, Javanese mysticism, Christianity like thread to weave the country into one unifying whole.

GLOBAL INSTABILITY

By the time the United States became involved in Indonesia, it had implemented a reverse course policy that was considered to be an important victory for Western powers in the Cold War. It eliminated the threat of a communist Indonesia and diluted the strength of Sukarno's anti-Malaysia message to the world. As Easter (2005, 57) notes, "By the summer of 1965 there was a consensus amongst Britain, the United States, Australia and Malaysia that Sukarno was an implacable enemy, threatening the stability of the region and leading his country to communism." The United States feared that Sukarno's death would push Indonesia directly into the arms of communism.

As Easter notes,

> The Western powers responded to this threat in a similar way: by using propaganda and covert action. British and Australian soldiers secretly crossed the jungle border to attack guerrilla units inside Indonesia and Britain and Malaysia gave aid to rebel groups in the outer Indonesian islands of Sumatra and Sulawesi. In addition, the British and Malaysians used covert propaganda to erode support for Confrontation and encourage disunity in Indonesia. (2005, 57)

INTERNAL CONFLICT

Internal conflict within Indonesia at this time could be described as an ethnic bloodbath. Sukarno was steadily moving Indonesia to the left toward adopting communist ideology, and the rift between the staunchly anti-communist army and the large Indonesian community party was growing wider. Sukarno's vision for unity based on Marxist and Islamic beliefs was miscommunicated to many as an opportunity to gain justice and initiate violent retaliation towards others who they believed had done them wrong.

Easter describes the violence during that period,

> The killings were not just political acts in the Cold War, they were also a complex sociological phenomenon and the perpetrators had a wide variety of local motives. The PKI had supported land reform in rural areas and this had created bitter resentment between peasant party members and small landlords. Muslims and, in Bali, Hindus were driven by religious fervor to slaughter the atheist communists. The killings sometimes had racial overtones, such as attacks on ethnic Chinese in North Sumatra. In the frenzy of violence people saw a chance to satisfy personal vendettas. Factors other than propaganda drove civilians to murder suspected communists. The killings were not a reaction to Western propaganda—they were the culmination of years of built up tension and hatred. (2005, 68)

It was during this time that the CIA decided to initiate a plot to assassinate Sukarno. Richard Bissell, a former CIA Deputy Director of Plans, testified that "there was discussion within the Agency of the possibility of an attempt on the life of President Achmed Sukarno of Indonesia which progressed as far as the identification of an asset who it was felt might be recruited for this purpose" (*Summary of Facts* 1975).

Bissell adds, "The plan was never reached, was never perfected to the point where it seemed feasible . . . the Agency had absolutely nothing to do with the death of Sukarno" (Report on CIA Assassination Plots of the Richard B. Cheney Files at the Gerald R. Ford Presidential Library, dated June 5, 1975). In reactionary mode, Sukarno banned rival political parties from government and opened the portal even more for communist activities. The media was used to spread Marxist and anti-Western propaganda. By this time, Sukarno was sixty-four years old and was suffering from health issues. It was well-known to many that his plan was to hand over presidential power to the communist PKI who would, in turn, control Indonesia after his passing.

The realization of Sukarno's vision would have been a failure for the United States. Indonesia was at a location of geo-strategic significance with access to vital sea lanes, with an abundant amount of natural resources and a population that exceeded 103 million people at the time. Easter comments, "The loss of Indonesia would also outflank American efforts to contain communism in South Vietnam" (2005, 56). As Frederick and Worden note,

> By 1965 Indonesia had become a dangerous cocktail of social and political antagonisms. The PKI's rapid growth aroused the hostility of Islamic groups and the military. One of the most serious points of contention was the PKI's desire to establish a force of armed peasants and workers in conjunction with the

> four branches of the regular armed forces. Many officers were bitterly hostile, especially after Chinese premier Zhou Enlai offered to supply the "fifth force" with arms. Sukarno' collapse at a speech and rumors that he was dying also added to the atmosphere of instability. (1993, 12)

REGIME TYPE

Indonesia's regime type could be portrayed as a Wayang theater production. Two political cultures played an influential role in governance: the Javanese aristocrats and the Islamic entrepreneurs. Adding even more complexity to the equation is the addition of ethnic, cultural, and economic cleavages as well as dissention from Indonesia's outer lying islands. Within the constitution, there are six components of state.

Sukarno saw an Indonesia operating under its own unique form of democracy. In an essay he wrote shortly after being released from prison, he articulates his vision for a new guided democracy: "Our democracy . . . must not be a democracy a la Europe and America, which is just a portrait of parties, a political democracy only . . . but a political and democratic democracy which gives one hundred percent sovereignty to the people in the ordering of their political and economic affairs."[1] Calling on his self-appointed prestige as leader and visionary of Indonesia's unity and independence, Sukarno proceeded to articulate the key components of his guided democracy initiative.

The US role in the 1958 Indonesian coup has been downplayed to a significant degree. Very little information can be found regarding details of Washington's role. It is assessed that the hasty successes of covert action in Iran and Guatemala may have created some policy "speed bumps" as to how these actions are planned and executed. As Stiefler notes, "The CIA's attitude towards covert action has vacillated over time. From 1947–1965, during what I call the 'honeymoon' period, CIA leadership could be counted on to enthusiastically support covert action in a variety of forms" (2004, 634).

The US role in covert action was downplayed because, unlike the cases of Iran and Guatemala, the United States had failed to influence the Sukarno government. In this case study, the influence of public dissent and leadership characteristics were more powerful than the other variables of global instability and regime change. US failure to mobilize the masses, despite the use of propaganda, shows that Sukarno's leadership and intense desire to unify the country were influential factors. Internal conflict, as translated by the failure to mobilize anti-Sukarno political parties that would run against Sukarno or gain seats in political institutions, such as what the PKI did, would also prove to be another powerful force. The lesson to be learned is that Sukarno defeated the CIA attempt to oust him because, despite his faults, he had still earned the trust and faith of his fellow countrymen and, indeed, exhibited a "cult of personality." He had an innate ability to relate to people of all classes, and he identified himself with the cause of Indonesian independence. He viewed himself as a revolutionary leader, and people believed in him. However, once ousted from power, Sukarno spent the rest of his lifetime living in seclusion, ostracized by Indonesians, and died from kidney failure.

Frederick and Worden note,

> The international scene was, for Sukarno, a gigantic stage upon which a dramatic confrontation between (as he termed them) the New Emerging Forces and Old Established Forces was played out in the manner of the Wayang contest between the virtuous and the evil. (1993, 10)

Overall, this case study analyzed the effects of five causal conditions on the fate of leaders. These causal conditions are leadership characteristics, public dissent, global instability, internal conflict, and regime type. The QCA process intends to establish characteristics of each factor that assists the research in assigning the degree of membership of each case with the causal factor while taking into consideration the within-case context of each case that influenced the individual degree of membership.

Sukarno was a member of the minority population, he had a college degree, and he was under the age of forty years old when he assumed power. Public dissent is divided, with more than half of the population supporting him and several political groups wanting to oust him from power. He assumed power during a tumultuous time when Islamic influences were becoming stronger within the country and, in regard to global instability, the world was greatly divided between democracy and communism. With respect to internal conflict, the country was in a state of crisis at the time of the covert action event. While there was no coup, revolution, or regime change within five years prior, the country was undergoing a cultural shift between religious groups as well as growing dissent from inhabitants of outlying islands. Considering regime type, Indonesia was attempting to create a system based on democracy and communism. Sukarno called this new authoritarian regime "guided democracy" with him as the guide.

The foundation of QCA is the truth table, which allows the researcher to list and analyze many possible combinations of variables prior to arriving at the outcome. Ragin and Rihoux outline the steps to creating a truth table: "First, construct the truth table based on theoretical and case study information. Second, analyze the truth table. Finally, take the results back to the cases and theory to evaluate the findings" (2004, 4). Employing fsQCA, this researcher assessed how various constellations of leader characteristics, public dissent, global instability, internal conflict, and regime type affected leader survival after covert action events.

NOTE

1. A selection of his writings from this period appears in Sukarno, *Dibawah Bendera Revolusi*.

Chapter Five

Dominican Republic, 1959–1961

The decision of the CIA to leverage the dissident community in order to overthrow President Trujillo in May 1960 was a watershed moment in the history of US covert action. By providing funding to a small group of Dominicans, the United States was able to plan and execute the removal of a leader from power in an effort to prevent a country from falling under the influence of communist control. This strategy involved the training of a paramilitary force and equipping them with small machine guns and grenades, funding Dominican dissidents to operate radios that were used to spread anti-Trujillo messages to the civilian population, managing administrative tasks within government institutions, and conducting psychological operations in the form of leaflets and pamphlets drops over the Dominican Republic that were intended to exploit the excesses of the Trujillo regime.

A group of Dominican dissidents assassinated Trujillo on 30 May 1961. The duration of Trujillo's rule from 1930–1961 brought stability to the country; the nature of his rule was based on terror and brutality. It was during this time that civil liberties for the people were non-existent and human rights violations were common. Trujillo was also an egotist who, after assuming power, set out to recreate the country in his image by renaming cities and towns in his name and installing monuments in his image.

BACKGROUND

On the western side of the Atlantic Ocean lies a wide arc of land that was formerly formed by the coast of North America with roughly five thousand miles of shoreline. Its native inhabitants originated from what are now known as Cuba and Puerto Rico. Santo Domingo became the capital of a Spanish colony and formed the hub for commerce to and from neighboring islands. Soon thereafter, English, Dutch, and French colonists began to produce food crops, such as tobacco, cotton, and sugar.

Previously dominant foreign and domestic partners had all but been eliminated, thereby leaving the occupation government of the Dominican Republic to take on the large task of developing its infrastructure and commerce system. The trailblazers for this vast international development project were the United States Marine Corps, who essentially rebuilt the country while maintaining peacekeeping operations within the city and out in the rural hinterlands.

Calder writes,

> Marine guards patrolled the streets of Dominican cities, and Marine intelligence officers gathered information about those who opposed the occupation. When nationalist demonstrations occurred in Santo Domingo and Santiago during patriotic week in May 1920, the military governor jailed more than a score of writers for inciting unrest and asked for reinforcements; five hundred more marines arrived to bolster the existing force of two thousand. (1984, 17)

The Eastern District of the Marine command included the main sugar plantations, which prospered from expansion of the cane production business. Companies headquartered in the United States owned a large share of these plantations and benefited from the profits they produced. To protect them and the American assets, the US Marines recruited and trained Dominican soldiers to form the bulk of the security apparatus, also known as the Guardia Nacional. Fuller notes, "The Guardia Nacional remained an undisciplined organization with no commissioned Dominican officers" (Fuller, 1974).

LEADER CHARACTERISTICS

Rafael Leonidas Trujillo came from a modest, barely literate rural family; he was the third oldest among eleven brothers and sisters. His father, José Trujillo Valdez, was a postal clerk who was rumored to supplement his income with cattle herding. Trujillo was not considered well-educated, and at the age of sixteen, he earned a modest income as a telegraph operator and security guard at a sugar plantation. He was also involved in gang activity that specialized in theft and blackmail.

Roordin explains,

> While still a boy, he demonstrated the fascination with personal appearance, clothing and military decorations that suited him for a military career, earning

> the nickname "Chapita" for the battle caps he collected like medals. The year before his appointment to the academy, Second Lieutenant Trujillo had been acquitted of credible charges of rape and extortion by a commission of eight Marine officers. (1998, 21)

Trujillo's career path would have taken a different turn had he not taken the initiative to draft a calligraphed letter to the United States Marine Corps. In the letter, dated 18 December 1918, Trujillo requested (and was granted) a position with the Dominican National Guard. After becoming commissioned as an officer, it took Trujillo less than ten years to make brigadier general and commander in chief of the National Army. Trujillo held a stalwart admiration for the United States Marine Corps until his demise more than thirty years later. The United States initially supported Trujillo and viewed him as a protégé. As time passed, rumors of Trujillo's unbridled brutality quickly began to surface in conversations among policy makers in Washington, DC. Many of the US political elite were close and personal friends with high society Dominicans. Once firmly in power, Trujillo began to target the upper-class Dominican population by systematically eliminating them through false imprisonment and assassination.

NEUROTICISM

Neuroticism is a term used to describe an individual who is highly emotional, highly prone to anxiety, depression, violence, and anger. This trait is evident in Trujillo's self-proclaimed titles of Generalissimo Doctor Rafael Leonidas Trujillo; Benefactor of the Fatherland; High Chief Commander of Ground, Sea and Air Forces; Grand Patron of Arts and Letters; Father of the New Nation; among others. Scholars note that as Trujillo's power grew, so did his paranoia about his own security. He employed hundreds of spies and special researchers to pore over published periodicals to ensure that all negative press about him was eliminated from public scrutiny. The older he became, the more deranged and mentally unstable he appeared. His pursuit of wealth, power, and self-aggrandizement were veiled attempts to overcome his lower-class origins. He punished those around him who he considered high in status, wealthier, and more successful.

EXTRAVERSION

Extraversion is described as personality types that seek excitement and experiences beyond their comfort zone. Essentially, it is the quality of being confident and outgoing. Trujillo joined military service and underwent training with the US Marine Corps. Four years after successful completion of training, Trujillo graduated and rose to the ranks of lieutenant colonel. He made the rank of general officer shortly thereafter. His meteoric rise within the military ranks is largely attributed to his quest for more power and influence. The military was the fastest route to this objective and where he perfected his extroverted and narcissistic tendencies.

OPENNESS TO EXPERIENCE

This type of personality describes individuals who are bold and adventurous, dislike routine, prefer abstract thinking, and have a wide range of interests. The opposite of this personality trait is individuals who are traditional, non-artistic, and non-analytical. Trujillo's biographer Robert Crassweller discusses Trujillo's relationship with women in detail, noting,

> The Boss enjoyed thousands of conquests and groups of young girls were brought to his palatial residence two or three times a week for sexual orgies; the older "The Big One" grew, the younger his paramours. To cater to the Generalissimo's spectacularly carnal demands the National Palace maintained a fully staffed Office of Procurement whose agents scouted the Caribbean and the world in the search for new girls to entertain The Boss. (1966, 47)

Additionally, Trujillo secured his position by eliminating his opponents. His infamous death squads systematically located and eliminated opposition leaders or those who demonstrated any sign of resistance. His brutality had no limits. Methods employed included pushing victims off piers and leaving them to drown, using machetes to kill, and throwing children against trees. The brutal death squads executed most techniques as a means of eradicating the minority population. During this turbulent period, the United States held firm to its Good Neighbor Policy and chose to maintain a position of nonintervention. Under Trujillo's rule, the Dominican Republic was modernizing its sugar industry; creating a solid internal infrastructure of roads, schools, and hospitals; and liquidating the country's debt to a zero sum within his first seven years in office. It was also noted that his income was $200,000 a month, which equals close to $2,000,000 in 2018.

AGREEABLENESS

This personality trait describes individuals who are compassionate, trustworthy, lenient, and softhearted. In contrast, it can also describe individuals who are hostile, cynical, manipulative, vengeful, and ruthless. Trujillo often hosted visits from Mrs. Eleanor Roosevelt and other dignitaries. However, the relationship with the United States soured toward the end of his rein in power. In an act of vengefulness against a Dominican scholar and Columbia University professor who was outspoken about the regime, Trujillo ordered the kidnapping of Dr. Jesús de Galíndez from the United States and transferred him to the Dominican Republic, where he was tortured, killed, and buried in an unmarked grave. Others who played a part in the kidnapping, including US citizen Gerald Murphy, the US pilot who unknowingly transported the scholar to the Dominican Republic, also disappeared without a trace.

CONSCIENTIOUSNESS

The personality trait of conscientiousness describes individuals who are able to manage their impulsive behavior through persistence motivation. These individuals also prize perfectionism in their quest for success and status.

Diederich recalls in *Trujillo: Death of the Goat*,

> The cult of adulation had no limits. The capital city bore Trujillo's name, as did the highest mountain peak and hundreds of streets, parks and sites. The tropical landscape was literally dotted with his likeness in statues and on bronze plaques. Every town, sports event, or billiard game had to be dedicated to the Benefactor of the Fatherland and Father of the New Fatherland. The humblest dirt-floor hut carried the sign: "In this house Trujillo is Chief." In the streets of the capital, neon signs blinked the message "God and Trujillo." (1978, 27)

Roordin adds,

> Trujillo asserted complete authority over the rebuilding of the city, which he would rechristen Ciudad Trujillo in 1936, giving him the ability to reorient the social organization of its population to reflect his control. His reconstruction of Santo Domingo, including a new bridge over the Ozama and greatly improved port facilities, deepened the respect that many Marines and Navy officers already had for him, while his reconfiguration of Dominican society through grandiose personalism (symbolized by naming the new bridge, the new port, and the city itself for himself) heightened the diplomats' distaste for him. (1998, 58)

PUBLIC DISSENT

It is interesting to note that this is the first case study in which the dissident population had a tremendous influence on the life or death of a leader. Dissent is an important component of the Trujillo chapter. During this time, speaking out against the government guaranteed execution, torture, and certain death. Trujillo had created a corrupt government based on fear and punishment. His most loyal officers benefited from the power and prestige by taking wealth for themselves.

GLOBAL INSTABILITY

In December 1960, five months prior to the assassination of Trujillo, the CIA station chief at Santo Domingo, Henry Dearborn, described the local security situation as a police state run by the Trujillo regime. Formal diplomatic relations between the United States and the Dominican Republic did not exist at this time. However, the station chief was successful in developing a relationship with the Consul General who was "able to establish direct contact with leaders of various opposition groups whose sole aim at this time was to bring about the assassination of General Trujillo" (Roorda 1996).

From the declassified CIA letter:

> When opposition plans to attempt the assassination reached the stage that it would be attempted at any time, the chief of station arranged with an opposition leader for the passing of a coded telephone message which, when made would confirm that the attempt had been successful. When the assassination had been carried out successfully, the station, when it received the message, was able to verify the fact many hours before it was announced officially by Dominican authorities. Immediately following the assassination, the station obtained additional details from an informant and three other sources which had previously been developed by the station for the purpose of following up on such an eventuality.
>
> Following the assassination of Generalissimo Trujillo, in the Dominican Republic, the Agency began a program designed to aid and strengthens the development of viable, non-communist organizations to play an active role in the post-Trujillo political life of the country. Moderate organizations were composed largely, but not exclusively, of middle class elements. The majority of leaders are vigorous young men, well-intentioned, though at times politically immature. During the Trujillo era many were jailed and tortured. (Roorda 1996)

INTERNAL CONFLICT

This section analyzes the factors that contributed to internal conflict within the Dominican Republic from 1959–1961 and if these factors led to larger problems in the future. The principal questions to be addressed are twofold. First, what was the security environment like during this time? Second, did internal conflict result in the spawning of other problems following the 1961 covert action event?

The plan to kill President Trujillo, as laid out by the conspirators, was simple and direct. An action group of eight assassins deployed in three high-powered sedans was responsible for shooting the dictator. Once Trujillo was dead, General Ferdinand, the military minister and ranking army general, was to assume power in Santo Domingo.

REGIME TYPE

The US foreign policy of nonintervention allowed the Dominican Republic and its dictator to operate without oversight or accountability. Gross humanitarian violations were committed without consequences. The country was far from earning its Good Neighbor title. Roordin adds, "The connection between the Good Neighbor Policy and the Trujillo regime offers a revealing perspective on the debate over how democratic states should treat authoritarian governments in the United States to control or even to consistently influence the forces of nationalism and personalism that intertwine in the figure of a dictator, whatever he is called: Fuhrer, Duce, Chairman, First Secretary, or Generalissimo" (1998, 1).

According to Roordin,

> Dictatorship has been the greatest problem of the twentieth century, rising up after the Great War to fuse the worst aspects of the old autocracies to new ways of thinking and of mobilizing the masses. Gaining momentum with the global depression in the late 1920s and 1930s, the formation of dictatorships in Europe, Asia, and Latin America was a central challenge to the Hoover and Roosevelt administrations. (1998, 4)

These states essentially operated without discipline, oversight, or accountability. They were single-party political systems that fed off of hyper militarism, ethnic hatred, and political greed and corruption. Trujillo's dictatorship represented the antithesis of democratic principles espoused by the United States in its foreign policy.

He adds,

> The Good Neighbor Policy developed an effort to encourage order, solvency, cooperation, and liberalism in Latin America through persuasion and benefits, rather than by force, but it would be put to a variety of unforeseen uses by Latin American leaders, especially those with autocratic power like Trujillo. (Roordin 1998, 4)

CONCLUSION

An interesting pattern to note in the Dominican Republic case study is the link between leadership characteristics and public dissent. Trujillo was a brutal dictator who grew increasingly more deranged as he advanced in age and surrounded himself with intelligence officers and henchmen who ensured that any semblance of resistance was eliminated with vicious brutality. His strongest supporter was John Abbes, who was also his chief interrogator and mastermind of various torture techniques. Abbes had a notorious reputation for devising ways to make Trujillo's enemies suffer for their opposition to the regime. His death squad was responsible for the kidnapping, extortion, and elimination of thousands of Haitians and Dominicans. Trujillo even went so far as to kidnap an anti-regime professor from the streets of New York City. He had no limits to his power, and the United States disregarded this brutal regime. Another interesting association is the link between leader characteristics and global instability. This case study, more so than the others, effectively showcases US foreign policy toward Latin American countries. Trujillo and his staff feted American dignitaries during their visits to the Dominican Republic. The lengths the United States went through to establish the Dominican Republic as a good neighbor appeared almost excessive during that time. In addition, Trujillo was considered more of an annoyance by American policymakers than a true peer.

Roordin writes, "Curtis and Cabot were disdainful of Trujillo's lower class social status. Both were upper-class, white Easterners who favored the companionship of the traditionally dominant Dominican Republic, an oligarchy of light-skinned, well-educated elite families in Santiago, Puerto Plata, and Santo Domingo" (1998, 43).

The extent of the Trujillo regime's brutality was viewed skeptically by White House policymakers. It was not until friends of US policymakers, elite members of Dominican society, were kidnapped and tortured that the administration began to take notice. For the first time in this series of case studies, the concept of dissidents is introduced as an influential factor in determining a leader's fate.

According to declassified transcripts from the US Congress, Senate, Select Committee to Study Governmental Operations with Respect to Intelligence Activities, Intelligence Activities and the Rights of Americans, 94 Cong., 2d sess., 1976, 1, it was noted that "During the course of a cocktail party in the Dominican Republic, a leading dissident made a specific request to Ambassador Farland for a limited number of rifles with telescopic sights and the ambassador promised to pass on the request" (Farland Affidavit, 9/7/75, p. 1). According to the Church Committee Report, "Documents indicate that consideration was given within the CIA to airdropping rifles into the Dominican Republic." At a 21 June 1960 meeting with an officer of the CIA's Western Hemisphere Division, Ambassador Farland reportedly suggested possible sites for the drops (CIA memo, 6/21/60). Additionally, the report states, "The memorandum recommended that the arms be provided, since the fall of the Trujillo regime appeared inevitable, and therefore United States relations with the opposition should be as close as possible. Providing the arms as requested would contribute significantly toward this end" (Church Committee Report, 183, and CIA memo, 7/1/60). Rafael Trujillo came to power in the Dominican Republic in 1930 and, for most of his time in office, the United States supported him and considered him its protégé. However, as time passed, Trujillo's rule became increasingly dictatorial and cruel, which placed him in the crosshairs of the United States for further scrutiny (US Congress, Senate, Select Committee to Study Governmental Operations with Respect to Intelligence Activities, Intelligence Activities and the Rights of Americans, 94 Cong., 2d sess., 1976, 1).

Roordin explains,

> Trujillo's control of the Dominican military assured that stability would prevail under his rule, but sometimes stability did not live up to its billing. Although imposing order with devastating force, the regime's foreign and domestic policies did not coincide with US interests in several important regards. The Dominican Republic became a difficult place to do business, a querulous participant in negotiations, and a major cause of Caribbean disquiet, including genocide, war scares, and assassinations. (1998, 2)

The United States simply could not be seen as an associate of a Latin American dictator with a penchant for cruelty and brutality. This brutality and fear caused US policy makers to make parallel comparisons between Trujillo and Castro. Thus, plans were put into motion to remove Trujillo from power.

This case study analyzed the effects of five causal conditions on the fate of leaders to include leadership characteristics, public dissent, global instability, internal conflict, and regime type. With respect to leader characteristics, Trujillo was a member of the minority population, he did not have a college degree, and he was under the age of forty years old when he assumed power. Public dissent was a strong factor as most Dominican men by the age of forty had endured and survived Trujillo's reign of terror during which he eliminated men, women, and children on a whim. Minority members of society were massacred by the thousands, and people were kidnapped, tortured, and murdered. His affinity for women, as he was known for having four wives, countless mistresses, and thousands of sexual escapades with young girls, also created resentment within the machismo culture. Trujillo exploited the opportunity to use state funds to rebuild the country while claiming that the money came from his personal account. As pertains to global instability, the threat of communism was strong and heavily influenced foreign policy behavior of the United States. The United States was deeply mired in the Great Depression and considered the Dominican Republic to be a good neighbor and Trujillo to be its protégé as he had been vetted and trained by the United States Marine Corps.

With respect to internal conflict, the Dominican Republic was in a state of crisis at the time of the covert action event. The island was divided between the fairer-skinned sugar plantation owners and the black slave workers. As to regime type, the country was an autocracy run by an egotist and brutal dictator who became more sadistic and deranged in his old age.

Table 5.1.

Variable/ Membership	*Fully In*	*Mostly In*	*More In Than Out*	*More Out Than In*	*Mostly Out*	*Fully Out*
Leader Characteristics	X					
Public Dissent	X					
Global Instability			X			
Internal Conflict				X		
Regime Type					X	

Chapter Six

Leopoldville (Kinshasa), Democratic Republic of the Congo, 1960–1961

The story of the Congo is a tragic one. It is a country blessed with an extraordinary amount of natural resources, bountiful wealth, pride in its rich and colorful traditions, and infused with a strong spirit of independence. Yet, neither the country nor its people were ever free from colonial rule. Despite the efforts made to establish independence, the newly designated Congolese leaders used their power and authority to hoard their own wealth and create terror and fear against their own people, both as means to reinforce their own power. The brutality of King Leopold II, while cruel and vicious, was stark in comparison to the atrocities that the Congolese military and its leaders committed against innocent civilians. "Huntington uses the term praetorians, which specifically defines politicization of the military, to define the broad phenomenon of a politicized society. What differs about military involvement is that its coercive power 'trumps' all other political competition" (Barron 2013, 101). Many decades of civil war continued long after colonial rule disintegrated. The tragedy lies in the story of a vibrant and charismatic leader, Patrice Lumumba, who had a vision of independence for the Congo. He was a glimmer of light in the heart of darkness.

The Democratic Republic of the Congo is roughly the size of Mississippi and centrally located on the African continent. The Belgians discovered it in 1869 on a search for a missing missionary, David Livingstone, who was presumed lost somewhere in the vast spans of the Congo. Upon discovery of the country's abundant supply of natural resources, and Britain's rejection of adding Congo to its already existing colonies, Belgium took advantage of an opportunity to establish its own colony on the African continent. As Zeilig (2008, 63) notes, "Few had seen the written word before, and they were being asked to mark their X's to documents in a foreign language and in legalese. The idea of a treaty of friendship between two clans or villages was familiar;

the idea of signing over one's land to someone on the other side of the world was inconceivable."

As of 30 June 1960, the Congo and Central African Republics were French colonies; Uganda, Tanzania, and Northern Rhodesia were British colonies; Rwanda and Burundi were Belgian colonies; and Angola was a Portuguese colony. Also, at that time, the Congo produced 8 percent of the world's copper, 8 percent of its diamonds, 73 percent of its cobalt, and 60 percent of its uranium, gold, and zinc. Described by scholars as a place of extreme "dysfunctionalism," it seemed that the country never recovered from its colonial legacy under Belgian King Leopold II, who murdered millions of Congolese in his quest for natural resources for his personal gain.

In 1961, Undersecretary of State George Ball stated, "Should the Congo crumble into chaos and become a successful object of communist penetration, the Soviet Bloc will have acquired an asset without a price—a base of operations in the heart of Africa from which to spread its tentacles over this newest of continents" (Department of State Publication 7326 1961, 2). On 30 June 1960, under the leadership of its first democratically elected Prime Minister, Patrice Lumumba, the Congo was granted independence. Belgium immediately sent seventy-five thousand troops to protect its resources, thereby demonstrating to the world that the Congo was not truly independent. The United States' national security policy at that time, under the Dwight D. Eisenhower administration, was based on the containment of communism.

LEADER CHARACTERISTICS

Patrice Lumumba was born on 2 July 1925 in the Kasai Province, Belgian Congo. Lumumba's father was a farmer; he and had three brothers, one half-brother, and received his primary education from a Catholic missionary school. He was a complex man who had a voracious appetite for reading and a strong desire for self-improvement. He has been described as passionate, articulate, and intelligent but also very naïve. He saw endless potential for the independence of the Congolese people.

By 1960, King Leopold II's brutal reign on the colony had loosened, and the country was undergoing reform of its administrative processes, which included the creation of formal corporations owned by Belgium for the sole purpose of exploiting its natural resources. Additionally, Belgian and Congolese traditions had been grafted together, creating a hybrid culture of new customs based on the interweaving of Belgium and Congolese traditions. Lumumba had learned this hybrid culture and leveraged it to obtain popular support.

NEUROTICISM

Neuroticism is one of the most interesting categories of the personality model. It is characterized by an individual's outward demonstration of moodiness, anxiety, worry, envy, frustration, jealousy, and loneliness. Lumumba started out as an obedient postal clerk in Stanleyville. However, his demeanor began to change once he sensed the harsh racism and segregation that came with Belgian rule. In one instance, Lumumba disputed a salary discrepancy by writing a letter to the head of provincial finances. His polite query was answered with a vicious response, and Lumumba became more determined to stand against the perceived racism of his Belgian superiors. As Zeilig (2008, 398) noted, "From now on Lumumba would not be intimidated by his white bosses, nor would he accept their edicts if he regarded them as unjust. Lumumba had a growing sense of unease at the racism inherent in colonialism."

EXTRAVERSION

Extraversion, as defined by the Oxford English Dictionary, is described as an individual who is friendly and uninhibited, talkative with an energetic nature, whereas introverts are often described as solitary and reserved. A *New York Times* reporter wrote, "Tall, thin, intense, his eyes flashing behind his spectacles, Patrice Lumumba was a spellbinding orator who had created a nationalist party and had led it to victory in the Congo's first election" (*New York Times Magazine* 1981). Lumumba was certainly an extrovert with an outgoing and gregarious personality. He was a gifted orator and could relate to people from different social classes. In one of his earlier jobs as a beer distributor, he took this opportunity to sell Polar Beer to local bars and restaurants and, as a result, was able to mingle and network with people of all social classes.

Kuklick describes him as

> A flamboyant and charismatic leader, Lumumba had high hopes as a pan-Africanist for directing his new country into an honored place on the continent and in the world political community. But stability in the just-born country immediately broke down, and uncontrolled, the Congo's army mutinied and spread havoc rather than peace. (2014, 144)

Similarly, Zeilig notes,

> Lumumba also had to straddle two worlds, and sought acceptance in both. There was the rural one of his family in Onalua, who found him a bride, and insisted

> repeatedly that he stay with her. In addition there was the urban one and the prospect of social advance. Lumumba sought to negotiate both. He was also becoming a public figure, who participated in political and cultural discussions in the city. In managing these two, distinct universes, there were contradictions in his behavior. In the perfect and educated Congo that he envisaged in the 1950s he was the enlightened liberal, seeing women in a much more prominent position, even if they did not always play this role in his household. Lumumba had a keen sense of the struggle he faced to improve himself. This effort at self-education was a defining one. (2008, 459)

His first speech demonstrated his gift for public speaking and his ability to spark the fire of independence among the Congolese people:

> We are proud of this struggle, of tears, of fire, and of blood, to the depths of our being, for it was a noble and just struggle, and indispensable to put an end to the humiliating slavery which was imposed upon us by force. This was our fate for eighty years of colonial regime; our wounds are too fresh and too painful still for us to drive them from our memory. We have known harassing work, exacted in exchange for salaries which did not permit us to eat enough to drive away hunger, or to clothe ourselves, or to house ourselves decently, or to raise our children as creatures dear to us. We have seen our lands seized in the name of allegedly legal laws which in fact recognized only that might is right. The Republic of Congo has been proclaimed, and our country is now in the hands of its own children. (Van Lierde 1972)

OPENNESS TO EXPERIENCE

As Zeilig (2008, 211) notes, "From an early age Lumumba forged his own path. Before long he chose his own schools, made his own decisions, and fought his own battles. As an adolescent, he immediately stood out, as a popular and strong character. Friends and family remember him as curious and audacious, confident in his own abilities." He also had three wives and several lovers, all within a six-year period. According to his son, Francoise Lumumba, "[his father] discovered in the course of 1960 that not all Congolese had the same interpretation of independence, our 'brothers' were fighting for something completely different" (Zeilig 2008, 148). Zeilig adds, "Lumumba wanted true national liberation and participatory democracy. However, the Congolese elite were jealous and fearful of what that meant and conspired with the Western groomed political and economic classes to have him removed" (2008, 148). The United States feared that he was leaning too far toward the Marxist Socialist left. Considering that Lumumba was a prolific reader, he had very little knowledge about Marxist literature and ideology (Zeilig 2008, 148). His most significant vulnerability was himself and his innocent faith in independence for Congo, which was a region ripe with internal strife, corruption, and greed.

AGREEABLENESS

The movie *Lumumba*, produced by Jacques Bidou and Raoul Peck and premiering at the Cannes Film Festival in 2000, portrays the young leader as a charismatic, intelligent, and articulate man who had the confidence to challenge injustice. He was well-liked by the Congolese people, and they supported his path toward political success. The film was distributed to Belgium, France, Switzerland, the United States, and Canada with positive reception. Film critics mentioned that the documentary provided an accurate depiction of the leader's life based on his biography written by Ludo de Witte *The Assassination of Lumumba.* It was not presented as a propaganda film but more so as a documentary of the leader.

CONSCIENTIOUSNESS

Lumumba deeply understood the pain his people had endured under colonialism, which included torture and abuse under Belgian prison authorities. Lumumba's daughter Juliana commented in an interview,

> My father's political life lasted just three years, he developed with incredible speed. Everything that he finally demanded, immediate independence he was not the first to demand but he became the incarnation of independence; he became the representative and the symbol for immediate independence. He worked enormously hard, and I would watch him work, rewriting or practicing a speech. My father was a very affectionate person, when he was with us he always chatted and played . . . when he was there, he was really there. (Juliana Lumumba 2006)

PUBLIC DISSENT

By the early 1950s, a new class of Congolese intelligentsia arose from the population. This group consisted of Congolese men who were campaigning for national liberation and members of this group were actively involved in the struggle for independence. As Africa Scholar George Nzongola-Ntalaja notes, "It was a continuation of their fight for equality of opportunity in the colonial political economy where they experienced discrimination in respect to career and other economic opportunities, in addition to daily humiliation of colonial racism" (Nzongola-Ntalaja 2002, 31). Segmented tribal factions and a new social class created social tensions. "The Congolese elite possessed no capital of their own, major mining companies and businesses were owned by foreign companies. Most were employees themselves, as clerks and administrators in the colonial state. While the Congolese elite had a thirst for wealth, they were not involved in the ownership or production of it" (Ziegler 2008, 522). Many of the members of this newly established social class were born and raised under Belgian colonial rule. Many of these members undertook the requisite process to become enlightened and cultured in the way that was expected of them. As Lumumba's daughter recalls, the process to become an *evolué*, a French term to describe native Africans who were Europeanized through a process of education and assimilation and accepted European behaviors and values, was highly demeaning. Nevertheless, Lumumba and many others believed that it was the only path to opportunities for Congolese independence. It was merely a matter of time before they realized that this was not the case.

GLOBAL INSTABILITY

The breakdown of colonial ties in Africa left unstable and underdeveloped nation states vulnerable to Soviet influence. As a proxy war began between the United States and the Soviet Union over newly independent states without formal alignment, Congo was caught in the crossfire. Quick to establish a democratic state, the people of Congo elected their first president, Joseph Kasavubu, and prime minister, Patrice Lumumba. In the eyes of the United States, the Congo represented a beacon of light and there was much hope that the fledgling democracy would succeed. It was not long before members of the Eisenhower administration, including Secretary of State John Foster Dulles, Under Secretary of State Douglas Dillon, incoming Secretary of State Christian Herter, and American Ambassador to the United Nations Henry Cabot Lodge, became increasingly concerned that the Congo crisis would provide an opening for Soviet intervention. This concern led to an invitation to bring Lumumba to Washington, where he was regaled with the honors of a distinguished visitor. After the failed attempt to establish positive relations with Lumumba, the United States decided that it was better to remove the threat as opposed to letting it continue to exist. The result was a failed attempt to assassinate a foreign leader.

INTERNAL CONFLICT

Following the end of Belgian colonial rule, the Congo was left with a fledgling and fragile framework for governance, a new elite class of cultured Congolese, and a disgruntled army that ran amuck and became increasingly more powerful, violent, and corrupt as it grew. Described as "a vestige of premodern barbarism, populated by rapists, and bearing no resemblance to the world of modern armies" (Baaz and Stern 2011, 563) and "the ramshackle armed forces pulled together from the current warring factions" (Robinson 2012, 475), the Congolese army, also known as the *Forces Arme'es de la Re'publique De'mocratique au Congo* (FARDC), became the symbol of the country's highly corrupt, violent, dysfunctional, and decentralized military apparatus. Historically, the organization's modus operandi has emphasized exploitation rather than protection of the populace. As Robinson notes, "The sad truth is that the Congo/Zaire has never had any military tradition of its soldiers behaving properly, because its ethos has always been one of exploitation, not protection. The task of the original Belgian Congo's Force publique was to break all resistance to the economic exploitation of the state so that Leopold and his agents could plunder the territory" (Robinson 2012, 475).

The Congo's armed forces had never been trained to provide security or protection. Simply, its sole purpose was to extract resources on behalf of its patrons through a vast patronage network. As Robinson asserts, "The mere fact that the troops are armed is enough to overawe the populace; little more is required" (2012, 475). Its lackluster history dates back to 1888 when King Leopold II created the force as his private tool of resource extraction management. The *Force publique* was led by Belgian officers with Congolese natives forming the majority of the enlisted force. Lumumba's lightning-fast trajectory towards independence resulted in a haphazard construction of a Congolese army with natives in the officer ranks. The *Force publique* was renamed the *Arme'e Nationale Congolaise* (ANC), or Congolese National Army, but, as Robinson notes, "quickly degenerated into armed gangs of renegades often loyal to local ethnic leaders or secessionist regions rather than the central government" (2012, 477).

REGIME TYPE

Belgium's decision to grant independence so expeditiously was connected to political change in 1959, and it wanted to move quickly in hopes that, at the time of the election, there would be a Belgian ruling elite—80 percent of the Congolese parliament would be comprised of docile pro-Belgian members, with only 20 percent radicals. The Belgian government had planned for a protracted path toward democracy. However, internal strife and violence accelerated the process, which resulted in chaos. Unwilling to walk away from the vast wealth of the Congo region, Belgium was forced to assuage to the demands of the newly installed Congolese government. Lack of a disciplined Congolese military force and internal political corruption created an environment of fear and violence.

Lumumba was also pulled into the influence of external support. As a result of his sympathies, Soviet diplomats started spending money in support of Lumumba, pushing US President Dwight D. Eisenhower to the brink of requesting the deployment of North Atlantic Treaty Organization (NATO) forces. Lumumba also caused the Belgian business community to hire mercenaries to help the secessionist ambitions. By August 1960, the country was in disarray, and only a successful UN peacekeeping mission initiated by the Eisenhower administration would stabilize the Congo for five years after its independence. Contractors from the UN and the World Bank kept the public works and other services going in lieu of a stable Congolese political system.

The United States was concerned about Lumumba's public appeal and his leaning toward the Soviet Union. He had enlisted Moscow's help to ship Soviet airplanes, trucks, and technicians to the province where his support was strongest. In the final days, accounts of Lumumba's circumstances and death vary. The UN investigation concluded that he was killed on 17 January 1961. Coincidentally, President Eisenhower, in a nationally televised speech, addressed the American people for the last time as president on this day. The planning of operations involved only a handful of people: the Director of Central Intelligence, Allen Dulles; CIA Congo station chief Larry Devlin; and a CIA scientist who was asked to make preparations to assassinate or incapacitate an unspecified African leader. The CIA station chief testified that he had been ordered by the president to kill Lumumba. Assassination was intended to be conducted through biological materials (needles, rubber gloves, gauze masks), and it was to occur anytime and anywhere. The objective was to eliminate the threat through an injection into something Lumumba ate or used regularly for personal hygiene (e.g., food or a toothbrush). Shooting was also considered. The toxin was supposed to produce a disease that was indigenous to the Congo. On September 24, 1961, Dulles sent a cable to Leopoldville stating, "We wish to give every possible support in eliminating Lumumba from any possibility of resuming governmental position." (Church 1975). Lumumba was referred to as a political threat. CIA technicians developed and stockpiled toxins that could immobilize an individual for hours, days, or even months, or could kill a human being in a way that appeared to be the result of a deadly disease. "A formerly classified

CIA document notes that in November 1962, an agent (name redacted, but suspected to be Devlin) had advised Mr. Lyman Kirkpatrick that he had, at one time, been directed by Mr. Richard Bissell to assume responsibility, for a project involving the assassination of Patrice Lumumba" (Church Committee, 1975). Gerard and Kuklick add, "Regarding the US role at the end, they present a very strong circumstantial case. The United States, through the CIA, was demonstrably trying to do Lumumba in—directly and through its cooperating Congolese leaders—from August through November 1960." Barron adds, "Focus on the importance of the military, arguing that the weaknesses, rather than strengths, in the military's structure and function contribute to military coups and to the nature of the broader political context in which it exists" (2013, 101).

Barron suggests that "The structure and functions of the military contributed to and reflected the complex political, social, and economic turmoil of the period. Between independence and Mobutu's second coup, violence, instability, and foreign intervention were consistently integral to the Congo's political power struggles, in which Mobutu had, from the beginning, significant and growing influence" (2013, 112). This case demonstrated the influential power of leader characteristic, internal conflict, and regime type to a leader's survival. Had Lumumba been given the opportunity to make his dream of Congolese independence a reality, the world could have seen a very different turnout for this Central African country. Lumumba was intelligent, articulate, and passionate, but he grossly underestimated the greed, selfishness, and corruption of his country's elite leaders. The Congo case study demonstrates the association between leader characteristics, internal conflict, and regime type as captured in the study's model. The presence of each factor had a direct influence on the other, albeit in varying degrees.

One of the interesting associations to note in the Congo case study is the link between personality characteristics and internal conflict. While Lumumba enjoyed popularity among the Congolese people, his enemies from opposing regional factions were intent on his demise. It appears that no Congolese leader, no matter how educated, would stand a chance against opposing forces within the country and outside of the country.

This case also demonstrated the more influential power of regime type to a leader's survival. Had Lumumba and his league of educated peers had the foresight to understand the meaning of independence, they may have had a chance to see it succeed. The Congo case study does demonstrate the association between leader characteristics, internal conflict, and regime type as captured in the study. The presence of these factors has a direct influence on the other, albeit in varying degrees. Of all the case studies presented in this book, the story of the Congo is the most tragic. It is a story of the Congo's young and experienced politicians, a greedy Belgian government that refused to let go of its most prized colonial asset, and American leaders who unnecessarily influenced the history of a country in the most flawed and ill-prepared manner.

Finally, this case study analyzed the effects of five causal conditions on the fate of leaders to include leadership characteristics, public dissent, global instability, internal conflict, and regime type. Lumumba was a member of the minority population, did not have a college degree, and was under the age of forty years old when he assumed power. Public dissent was not as significant of a factor as he enjoyed a tremendous amount of public support from the Congolese people. He assumed power during a tumultuous time when the country was on the verge of gaining independence after enduring seventy-five years under Belgian colonialism. Moreover, internal strife ran rampant and tribal factions and independent political parties were asserting power and influence for control. With respect to global instability, the threat of communism was strong and heavily influenced foreign policy behavior of the United States. Regarding internal conflict, the country was embroiled in civil strife. A major military uprising occurred within days of Lumumba's assumption to power. As pertains to regime type, the Congolese people were passionate about gaining independence. They wanted to have control over their destiny.

This case study analyzed the selection criterion of leader survivability by using QCA and the latest data from the ARCHIGOS, World Bank World Development Indicator, and POLITY IV datasets. It proposes that leadership style and personality traits, combined with institutional strength, are dominant factors in assessing the survivability of leaders following a covert action event. It focuses both on leaders removed from power via covert action, as well as their replacements. It also proposes that leaders' age and level of education plays roles but are not always positive in terms of determining survivability. The secretive nature of covert action has generated enormous interest in the topic. However, there is no existing systematic attempt to research and analyze the selection criteria of the most important focus area, the survival of leaders affected by covert action events.

This chapter explored covert action events planned and carried out by the United States in the Democratic Republic of the Congo from 1960–1961. It focused on the political environment during that time and then discussed the leader characteristics of ousted pro-Soviet Prime Minister Patrice Lumumba, who had a good chance of returning to power prior to his assassination in 1961, and the effects of his leadership and policies on Congolese society. After discussing the overall aftermath of the covert action operation and how American forces used that action to influence the Congolese government, the chapter concluded with a detailed discussion of the degree of membership this case demonstrates for each factor presented in this study's QCA model. Finally, this case study concludes with a summary and an application of the findings into the degree of membership.

This case study highlights the importance of leader characteristics as well as the power of internal conflict in changing the course of history. Additionally, it highlights leader characteristics as a driving force behind survival. Moreover, regime type is also an influential factor in this case study, as the Congolese people's desire to create independence from Belgian colonialization was the key impetus for the events that took place during this time. This case study will continue the analysis of US-supported covert action events and determine the extent to which leader characteristics, public dissent, global instability, internal conflict, and regime type were influential factors in the leader's survival. It is interesting to note how additional independent variables might influence the results of this case study. One independent variable to consider for future research would be the introduction of technology and how it may have changed the course of events.

Table 6.1.

Variable/ Membership	*Fully In*	*Mostly In*	*More In Than Out*	*More Out Than In*	*Mostly Out*	*Fully Out*
Leader Characteristics	X					
Public Dissent					X	
Global Instability				X		
Internal Conflict	X					
Regime Type	X					

Chapter Seven

Cuba, 1961–1962

This chapter explores covert action events planned and carried out by the United States in Cuba from 1961–1962. It will begin with a brief overview of the political environment during that time and then discuss the leader characteristics of Fidel Castro and the effects of his leadership and policies on Cuban society. After discussing the overall aftermath of the covert action operation and explaining how and why American forces used that action to influence the Cuban government, the study will conduct a detailed analysis of the degree of membership this case demonstrates for each factor presented in this study's QCA model. Finally, this case study concludes with a summary and application of the findings into the degree of membership matrix. It highlights the importance of public dissent and internal conflict in determining a leader's fate.

BACKGROUND

The state manages and governs Cuba and its people. Repression comes in the form of a strict single-party communist state run by the now deceased Fidel Castro, and his brother, Raul Castro. Under their leadership, the country is managed and surveilled by a tight network of government spies who have a great amount of control over the media, education, business, and foreign relations. This inward society prompted interest from outsiders, especially when its new leader began the process of agricultural reform and nationalization of its natural resources.

Husain notes,

> The day before President-elect Kennedy's inauguration, then-President Dwight D. Eisenhower had briefed him about numerous foreign policy issues, including

> Cuba, and a secret US effort to train anti-Castro guerrillas in Guatemala. In advising Kennedy on what to do with the thorny issue of Cuba, Eisenhower had recommended that this effort be continued and accelerated. Three days later, at a State Department meeting on Cuba, the then CIA Director Allen Dulles reported that US Special Forces teams from Fort Bragg, North Carolina, had trained between five and six hundred Cubans in Retalhuleu, Guatemala, for an invasion of Cuba. Whereas the previous administration had authorized eventual airdrops of the trained guerrillas onto the island, Kennedy now faced the decision of whether to continue the secret operation. One of the dilemmas confronting the administration, Kennedy's advisers told him, was the difficulty of maintaining secrecy. (2005, 24)

Husain adds,

> The landing point at the Bay of Pigs was part of the deception. The site was a remote swampy area on the southern coast of Cuba, where a night landing might bring a force ashore against little resistance and help to hide any US involvement. The original invasion plan called for two air strikes against Cuban air bases. A 1,400-man invasion force would disembark under cover of darkness and launch a surprise attack. Paratroopers dropped in advance of the invasion would disrupt transportation and repel Cuban forces. Simultaneously, a smaller force would land on the east coast of Cuba to create confusion. (2005, 27)

Unfortunately, many of the air force bombers missed their targets and failed to destroy any of Castro's air force assets. Shortly thereafter, the media reported the news and the plan was thus no longer secret. Put simply, the invasion failed. The covert action strategy involved sabotage, psychological warfare, and intelligence collection.

Husain ventures even further to explain, "Secret US plots to assassinate Castro had been in place since 1960, and a total of at least eight were planned between 1960 and 1965. Some of the outlandish proposals included lacing Castro's shoes with thallium salts so that his beard would fall out, and poisoning the dictator's beloved cigars with botulinum" (2005, 27).

LEADER CHARACTERISTICS

In his article titled, "Fidel Castro and the Cuban Revolution" Husain notes,

> Fidel Castro Ruz was born on August 13, 1926, on his family's sugar plantation near Biran, Oriente Province. His father, an immigrant laborer from Spain, owned a 23-acre plantation. As a boy, Castro worked on his family's sugarcane fields. Unlike many young Cubans of that time, he received an excellent education at Jesuit institutions Colegio La Salle and Colegio Dolopes in Santiago and

> then Colegio Belen in Havana. He was voted the school's best athlete mainly based on his baseball prowess. Castro continued his education at the University of Havana, where he worked on a law degree. (Prevost 2007, 19)

Castro described his father, Don Angel, as "a man of great will and great determination. He taught himself to read and write, with great effort" (Ramonet 2007, 24). He recalled his mother Lina, a Cuban, as "practically illiterate, and, like my father, she learned to read and write practically on her own. An extraordinarily hard-working woman, she was a cook, a doctor, a caretaker for all of us" (Ramonet 2007, 30). Critics and scholars have described Castro as a murderer, an avenger with a martyr complex, a liar, a control freak, dangerous, and a powerful individual who had a vengeful streak. To others, he is viewed as a revolutionary, charismatic, gifted orator with a keen intellect and the ability to capture every detail.

NEUROTICISM

According to Judge et al. (2002), neuroticism is used to describe individuals who demonstrate high levels of anxiety, are prone to depression, insecure, hostile, and vulnerable.

A declassified CIA personality profile notes,

> Castro's egoism is his Achilles heel. The extreme narcissistic qualities of his personality are so evident as to suggest predictable patterns of action during both victory and defeat. When he is winning, he must control the situation himself without delegation of authority, and he must continue to seek new areas of authority to overthrow. When faced with defeat, his first concern is to retreat strategically to a place where he can regroup his assets and personally lead another rebellion. (Central Intelligence Agency, Psychiatric Staff, "Psychiatric Personality Study of Fidel Castro, 1961)

EXTRAVERSION

Judge et al. (2002) also point out that extraversion describes leaders who are sociable and assertive and exude positive energy. According to the CIA,

> Castro's aggressiveness stems from constant attempts to achieve a special position that is denied him. When he achieves what he desires, he needs constant reassurance that he is justified in occupying this special position. In the past, he has sought approval from varying sources but currently he is wringing it from the Cuban masses, the current source of his sense of power and prestige. As long

> as the masses continue to support him, he will not suffer from anxiety, depression, or overt psychiatric symptoms. The chronic threat to the equilibrium of his personality is that this source of gratification might be withdrawn. (Central Intelligence Agency, Psychiatric Staff, "Psychiatric Personality Study of Fidel Castro, 1961)

Ramonet recalls, "He is a leader, who lives, so far as I could see, modestly, austerely, in almost Spartan conditions: there is no need for luxury; his furniture is sober; his food is frugal, healthy, and macrobiotic. His are the habits of a soldier-monk" (2007, 11).

Guerra explains,

> Brilliant and baroque, Fidel Castro has a visceral need to communicate with the public. He is aware that one of his main qualities is his ability to speak—to convince and persuade. He is better than anyone at capturing an audience's attention, holding them in his power, subjugating them, electrifying them, and eliciting storms of applause. There is no spectacle that can compare with Fidel Castro giving a speech. Standing, his body swaying, his hand often on the microphone, his voice thunders, his eyes fix on the crowd, he pauses . . . then moves his arms like a bronco-rider, raises his index finger and points at the suddenly tame crowd. (2007, 13)

OPENNESS TO EXPERIENCE

This leadership trait describes individuals who are curious about the world and who seek knowledge at different levels. Castro, for example,

> Castro is an ideal revolutionary leader, agitator, and fomenter of unrest, but he has no capabilities for organization and administration nor does he have any concern for the implementation of detailed plans. Furthermore, he can trust no one sufficiently to enable him to delegate authority. (Central Intelligence Agency, Psychiatric Staff, "Psychiatric Personality Study of Fidel Castro, 1961)

AGREEABLENESS

According to a declassified CIA personality profile,

> Castro has a constant need to rebel, to find an adversary, and to extend his personal power by overthrowing existing authority. When his self-concept is slightly disrupted by criticism, he becomes as emotionally unstable as to lose to some degree his contact with reality. If significant vulnerable aspects of his personality were consistently attacked by those he now looks to for approval,

> the result could be personality disorganization and ineffectuality—possibly even clinical emotional illness. This illness would probably be depression or some variant of depression, such as an overexcited state, an addiction, or an increase in suspicion to the point of complete withdrawal from reality. (Central Intelligence Agency, Psychiatric Staff, "Psychiatric Personality Study of Fidel Castro, 1961)

CONSCIENTIOUSNESS

Ramonet notes, "Castro likes precision, accuracy, exactitude, punctuality. Whatever the subject he may be talking about, he does mathematical calculations at amazing speed. No approximations for Fidel. He remembers the smallest detail." (2007, 13) According to a declassified CIA personality profile,

> Additional sources of gratification and ego bolstering appear to be his relationship to Che Guevara and his brother Raul. There are strong indications that Castro is dependent and submissive to Che intellectually and that his emotional stability would suffer if Che did not maintain a steady, positive attitude toward him. Disruption of this relationship, therefore, would discomfit Castro and reduce his effectiveness.
>
> Paradoxically, Castro seems to be basically a passive individual who defends himself against his fears of passivity by overreacting in aggressive and sadistic ways. His over-activity, avoidance of routine, lack of organization, impulsiveness, temper tantrums, and masochistic tendencies (including a wish for martyrdom) appear related to passive feminine wishes or identification. His compulsive need to be "on top" and never to submit to control or authority is another indication of his fears regarding passivity.
>
> Castro's consistent pattern of strong insistence on undoing the wrongs of "the little people," his preoccupation with the care and feeding of the poor masses, his concern for educational opportunities for the underprivileged, and his wish to be known to them as a benevolent "brother" indicate that he is to some degree conscience-stricken. His extreme punitive measures against rape and theft also indicate a backlog of unconscious guilt, which may be exploited to his disadvantage. (Central Intelligence Agency, Psychiatric Staff, "Psychiatric Personality Study of Fidel Castro, 1961)

The personality profile concludes with the following assessment,

> Although he depends on the masses for support, he has no real regard for them and does not trust them sufficiently to hold elections. His first consideration is to maintain power control for himself. He probably would destroy both himself and the Cuban people to preserve this status. This is the basis for his continuing

> the revolutionary stage beyond its period of usefulness. (Central Intelligence Agency, Psychiatric Staff, "Psychiatric Personality Study of Fidel Castro, 1961)

A leader born from a wealthy family, which began with an immigrant laborer from Spain and illiterate mother who owned a sugar plantation, took control following an era of repression and violence. His passion and faith for communist ideology led him to create a course of history for Cuba that initially created positive success for its people but ultimately ended in a dictatorship that controlled every aspect of civil society.

PUBLIC DISSENT

Following the fall of Cuban dictator Fulgencio Batista, Cuba was left with a politically fragmented society and a deep collective memory of violence, fear, and repression. It was during this time that Castro assumed leadership. As Guerra notes, "Unlike previous politicians, Fidel's rapid-fire vocal delivery paralleled the legislative velocity of his government. Creating agencies of state power and issuing laws by decree, the revolutionaries swept in a new age of ethical, activist government within weeks of victory" (2012, 45). Within one year of his assumption to power, Cuba's economy experienced positive changes. While prices fell and salaries increased, there was still a drastic difference between the luxuries experienced by urban dwellers and those who lived in the rural hinterlands. The former enjoyed modern luxuries such as indoor plumbing and running water, whereas the latter did not.

According to the CIA Weekly Current Intelligence Report dated 22 December 1960,

> The Cuban opposition to Castro at present includes large number groups both inside and outside Cuba ranging from organizations of former Batista associates to those of disillusioned ex-supporters of Castro. While the opposition has grown rapidly in recent months, it is not at this time a serious threat to Castro's position. (1960, 1)

> A further handicap has been the failure of much of the opposition to come up with a well-defined political and economic program, which would have sufficient appeal to win support among Castro's present backers. Those groups having adequate programs lack the means of communicating them to the Cuban people. Despite reports of increasing discontent among peasants and other low-income groups—the broad base on which Castro's power was built—such disenchantment has yet to become critical. (1960, 2–3)

In sum, while there were ample numbers of dissident groups in existence during Castro's reign, very few had effective means to communicate their platform to the Cuban people. Despite sharing common ground through their support for Batista, the groups did not have a singular, cohesive political platform from which to develop their message. That, combined with Castro's tight hold on all workings of Cuban society, made the dissemination of information difficult.

GLOBAL INSTABILITY

Cuba's charismatic and very socialist leader, Fidel Castro, quickly dashed any notion of a peaceful democratic transition. He believed passionately that any change that would come to Cuba would be determined and implemented from within the country. Despite commercial, economic, and financial embargoes led by the United States in the early 1960s, Castro remained stalwart in his vision for the country. Moreover, influence from the international community through political means did nothing to deter the domestic politics of the small island country.

By May 1959 when Fidel signed the long-awaited national Law of Agrarian Reform, complaints from US investors flooded the US Embassy in Havana. With massive sugar plantations and other major US investments now at stake, the same unsubstantiated charges of incipient communism that Batista had used against all opponents acquired renewed currency in the Eisenhower administration, some of whose top members were major stockholders in the UFCO, Cuba's largest landholder.

According to an intelligence report, "there was great concern on the part of American leadership about the presence of a government so closely aligned with the Soviet Union barely ninety miles from American shores" (Box 7, folder "Intelligence - Report on CIA Assassination Plots [1]" of the Richard B. Cheney Files at the Gerald R. Ford Presidential Library). A Memorandum for the Record dated May 15, 1975, states,

> During the period 1959–1961: a number of documents were found that referred to the intent on the part of some of the Cuban teams operating inside Cuba to attempt to assassinate Castro. Additionally, the files indicate that there was considerable interest in procurement of an air plane and air drop of Springfield rifles with telescopic sights. Phase I Plans involved the preparation of poison botulism pills by the CIA; the delivery of those pills to organized crime figures that in turn were to get the pills delivered to contacts they had in Cuba for placement in a beverage to be drunk by Premier Castro. Participants included Richard Bissell, Deputy Director of Plans and Colonel Sheffield Edwards, the Director

of the Office of Security of the CIA. (Declassified CIA Memorandum for the Record dated May 15, 1975)

The following facts primarily come from interrogation of Colonel Edwards and the Phase I case officer: the particular means by which the plans were to be carried out was by placing botulism pills in his food. The syndicate personnel were to get the pills to a person in Cuba who would have access to Castro. Supposedly, the reason pills were used was that the syndicate personnel could not recruit personnel to undertake the assassination through gunfire because the chance of survival and escape was small. However, the case officer felt that it was the Agency itself that desired to have pills used rather than gunfire. Colonel Edwards said that he, himself, checked out the pills on some guinea pigs because I wanted to be sure they worked.

Fidel Castro ceased visiting that particular restaurant at approximately the same time the pills purportedly arrived. The case officer said it was his recollection that there were two passages of pills to Cuba, the second one which was made to someone who had access to a restaurant which Castro frequented. After the second attempt failed, the case officer said the pills were returned to the CIA (Box 7, folder "Intelligence - Report on CIA Assassination Plots [1]" of the Richard B. Cheney Files at the Gerald R. Ford Presidential Library).

INTERNAL CONFLICT

As Guerra describes,

> By any measure, life in Cuba during last two years of the Batista regime can be characterized as controlled chaos. In addition to suffering economic recession, massive graft at all levels of government, and the takeover of the most profitable sectors of the tourist trade by the US mafia, Cubans contended with a political context of violence that they did not fully comprehend until January 1959. Behind years of fragmented reporting and contradictory accounts was a full-fledged, much bloodier civil war than most had realized. (2012, 39)

While Castro promised better conditions and an improved economy for the people, his vision changed during his tenure to manipulation and eventual termination of opponents within his regime. He believed in a united country under communist rule. Despite his Spartan lifestyle, he still enjoyed many luxuries that were not afforded to the masses. Some examples include palatial seaside homes rented to him by wealthy Cuban landowners for a mere one dollar. While other Latin American countries such as Guatemala and Chile were transitioning away from totalitarian rule, Castro enforced his draconian policies even harder on the people. The Cuban constitution included verbiage on state-sanctioned repression and described punishments for those who disobeyed. During his reign, virtually all civil and political rights were stripped from the people. While schools were built and children were afforded education, the curriculum was only centered on Marxist socialist ideology. There was no room for independent thinking in a repressive, collective society bound together in a right network of surveillance and spies.

REGIME TYPE

Cuba is a one-party communist state. Castro's control over every aspect of the country and its citizens created a repressive environment that operated in a power vacuum. Individual freedoms were ignored and human rights violations were common.

CONCLUSION

Castro and the Cuban revolution survived many decades because it was fueled by the people's passion for nationalism and independence. Through fear and control, Castro wielded his influence over the people and infused elements of Cuban society with his brand of socialism. The Cuba case study highlights the significance between leader characteristics and public dissent. The presence of these two primary factors has a direct influence on the other, and, ultimately, on the fate of the leader. Finally, this case study analyzed the effects of five causal conditions on the fate of leaders, to include leadership characteristics, public dissent, global instability, internal conflict, and regime type. With respect to leader characteristics, Castro was not a member of the minority population, he did hold a college degree, and was under the age of forty years old when he assumed power. Public dissent was an insignificant factor, as dissenters lacked an organized platform and failed to distribute their message effectively to the people. Castro was viewed as an egotist, as well as a charismatic and a gifted orator. He was born into wealth but instead chose to embrace a Spartan lifestyle based on minimalism. He embraced communist ideology and infused it into every aspect of Cuban society. He assumed power during a tumultuous time when the country had ended a brutal dictatorship under Batiste. Regarding global instability, the threat of communism was strong and heavily influenced foreign policy behavior of the United States. In reference to internal conflict, the country was undergoing upheaval with a leader who was socialist and communist. Communism is typically identified as a political system, whereas socialism is an economic system. As pertains to regime type, there were no aspirations to become a democracy nor did its new leader envision any such plans. The leader whose battle cry was "Socialism or Death!" passed away at the age of ninety on 26 November 2016.

Table 7.1.

Variable/ Membership	*Fully In*	*Mostly In*	*More In Than Out*	*More Out Than In*	*Mostly Out*	*Fully Out*
Leader Characteristics	X					
Public Dissent		X				
Global Instability	X					
Internal Conflict				X		
Regime Type			X			

Chapter Eight

Republic of Vietnam, May–November 1963

In the classic poem "Sailing to Byzantium," poet William Butler Yeats opines about the tragedy of old age and describes a passionate desire to escape into a dreamland full of beauty. The lesson learned in this case study is that the gift of old age is not always a guarantee, as seen in case studies focusing on episodes in which leaders were assassinated at an early age.

This chapter explores covert action events planned and carried out by the United States in the Republic of Vietnam in 1963. It will begin with a brief overview of the political environment during that time and then discuss the leader characteristics of President Ngo Dinh Diem, who had a good chance of returning to power prior to his assassination in 1963, and the effects of his leadership and policies on Vietnamese society. After discussing the overall aftermath of the covert action operation and how and why American forces used that action to influence the Vietnamese government, the study will conduct a detailed analysis of the degree of membership this case demonstrates for each factor presented in the QCA model. Finally, this case study concludes with a summary and application of the findings into the degree of membership matrix. It highlights the importance of public dissent and internal conflict in determining a leader's fate.

BACKGROUND

At the time of the covert action event under consideration, Vietnam, a longtime French colony, was still adjusting to its newly established role as an independent nation. Previously, it had endured a prolonged battle with its French colonial masters that played out in the form of a rebellion, which resorted to guerilla warfare as well as conventional military strategies. Unable to hold on to its power in the country, France relinquished control and withdrew its forces in 1941. Many years of colonialism resulted in the creation of strong nationalist movements that were supported by Vietnam's Chinese neighbors to the north. These movements played a key role in leader survival.

Extending from the Chinese border to the Gulf of Siam, Vietnam stretches out roughly one thousand miles long, with its physical geographic characteristics resembling two large breadbaskets held together by a slender pole. This physical separation also manifested itself into political differences between the people's democracy, allied with communist forces on one hand and a pro-Western autocratic republic on the other. Dense forest and predominantly mountainous combined with rich land for rice cultivation, Vietnam had one common language: Vietnamese. With the exception of a small minority population comprised of Thai, Khmer, and Hmong members who were deeply antagonistic of the Vietnamese, the majority of the population lived in rural areas and harvested rice as their primary means of sustenance.

Political groups and parties were organized after the French took control, and many of those actors received support from China and Japan. These groups, primarily the Vietnamese Nationalist Party founded by Nguyen Ai Quoc (also known as Ho Chi Minh), implemented change through terrorist techniques. They had no primary platform except to oppose the existing regime. It was not until later that the groups integrated political objectives into their platforms, which ultimately gave them more strength and influence against the opposition.

While the French were efficient in running the country, they also created significant cultural and economic disparities between themselves and the Vietnamese people. Specifically, the French enjoyed elite positions in government, opulent living conditions, and high-quality healthcare, but opportunities for education, medical aid, employment, and basic housing and sustenance were denied to the local population. This disparity created a deep, antagonistic relationship between the two cultures. As uprisings and social movements began to gain momentum, the French responded with immediate force by imprisoning and executing dissenters. Violence fueled the motivation of nationalist groups and, soon, these groups would reemerge with a political purpose. This new platform gave them greater influence and control against the opposition.

Ho Chi Minh was considered a visionary leader with Marxist-Leninist orientations. He was a loyal communist who played a key role in shaping the nationalist party in Vietnam. General Vho Ngyen Giap, Vietnam's lauded military revolutionary, was also a loyal communist who had been educated in China, earned a doctorate in Chinese history, and was well-versed in Mao Tse Tung's theory of guerilla warfare. These two leaders laid the foundation for the future of the country's political path. By 1951, the Vietnam Worker's Party was firmly an influential engine behind the fledgling Republic of Vietnam. It was also during this time that villages were organized into local, regional, and national hierarchies. Not long after, a military force was created utilizing the same template. The sheer will and determination of Minh and Giap, along with military aid and equipment from communist China and support from the peasant class, resulted in an overwhelming victory against the French in 1954. Support from the local populace was gained primarily through a propaganda campaign of psychological operations and information warfare.

LEADER CHARACTERISTICS

Born in the northern region of the Qang Binh Province, Ngo Dinh Diem was a member of one of Vietnam's elite families with Mandarin Chinese roots. His father was a court minister and Diem grew up with a devout Roman Catholic faith with a solid foundation in Confucianism. He had a strong work ethic and demonstrated steadfastness in his commitment to education. As biographer Shaw (2015, 25) noted, "Diem's exceptional determination, which did not alter in the face of adversity, became the mark of the man's later political life." Growing up, his parents hosted political dignitaries in their home, giving Diem constant exposure to many of Vietnam's visionary leaders.

According to Shaw, "Diem's father was unshakable in his opposition to violence or bloodshed. He stressed the revolution must come only through education. When the Vietnamese people were ready to look after their affairs, he argued, Vietnam would gain its independence from France naturally, with no need for killing" (2015, 25).

Diem embraced the Confucian teachings and demonstrated commitment to his studies at a very young age. He attended college and earned high marks as a dedicated student.

As Shaw explains, "He earned such high marks in the final examinations that the French offered him a scholarship to attend university in Paris. But Diem's great desire was to serve the Vietnamese people, and he turned down the French offer, choosing instead to stay in Vietnam, where he continued to excel academically" (2015, 27).

Scholars such as Jones (2003, 9) describe Diem as "cold, aloof, and nepotistic. He never revealed any propensity for democracy and preferred to keep power in the hands of himself and his family." This perception of the Vietnamese leader ultimately created a cultural divide between him and US advisors. As Shaw (2015, 27) emphasizes, "Diem's sturdy emphasis on religious, familial, and social duties placed a spiritual, moral and intellectual gulf between him and many of his American advisors."

Shaw adds,

> Diem led a disciplined life, and this was something the Vietnamese people revered because they believed that the quality of a man was determined by his ability to withstand hardship. Diem's scholarly, monk-like personality made him far more attractive to the Vietnamese people as a leader than, for example, a Western-styled, big-toothed, glad-handing, baby-kissing politician. Indeed, such a politician offended Vietnamese sensibilities, and yet it was this very model that many American advisors tried to push on Diem. (2015, 28)

Diem's stalwart sense of discipline and strong reputation for virtue made him a strong opponent to Ho Chi Minh. As a faithful Catholic in a predominantly Buddhist society, Diem did not have the foundation to refute religious differences. However, he concentrated his studies on Marxism and communism as a way to better understand his opponents' ideology. He was known for his administrative talents, which showcased his role within the local government. He was known for his disdain of Minh, who murdered Diem's brother by burying him alive. Minh also raided the Ngo Dinh home and destroyed Diem's collection of valuable books. Even after his assassination, Diem was revered by the Vietnamese people and his enemies for being steadfast in his faith as well as his honesty. He was also admired for standing firm against French colonialists, a trait that garnered loyalty from the people of Vietnam.

Shaw goes even further to explain,

> Diem believed that the individual needed to submit his will and talents to the greater good of the family, the community, and the nation. In his political philosophy, the individual did not have a right to political activity that threatened the downfall of a legitimate government. Diem's Catholic faith and Confucian principles were so robustly integrated as to make his political philosophy all but impermeable to contrary argument. (2015, 27)

NEUROTICISM

According to Judge et al. (2002), neuroticism is used to describe individuals who demonstrate high levels of anxiety, are prone to depression, insecure, hostile, and vulnerable. Diem was often described as cool and aloof by American advisors. He stood firmly by his Confucian faith and displayed stoicism to the public, even while experiencing tremendous hardship, believing that the traditional family and cultural ways of the Vietnamese should be protected from new, modernized changes.

As Jones notes,

> Diem personified a complex mixture of good and bad. He was a man of the highest moral principles, of strong will and, above all, a man who never panics, fully confident that he is the invisible hand of the Lord in everything he does. But he also had political blind spots that resulted from great caution, monumental stubbornness and equally monumental prejudices. He believed that the struggle against the Communists was purely military and that his government would suffer from a broadened political base. (2003, 33–34)

EXTRAVERSION

Judge et al. (2002) also note that extraversion describes leaders who are sociable and assertive and who exude positive energy.

Shaw argues,

> Diem had formed a pattern of going in and out of worldly affairs, a pattern he would continue until his death. When troubled, under threat, or just in need of quiet and reflection, he would retreat into a monastery or a similar religious institution. After his soul had been calmed and his thoughts organized through Christian contemplation, he would return to the hurly-burly of politics once more. (2015, 37)

OPENNESS TO EXPERIENCE

This leadership trait describes individuals who are curious about the world and seek knowledge at different levels. Diem demonstrated an ability to open his mind to new avenues of knowledge by studying Marxist and communist ideology as a university student.

AGREEABLENESS

The characteristic of agreeableness is used to describe a leader who conforms, is trusting of others, nurturing, and accepting (Judge et al., 2002).

Additionally, Shaw writes, "A good friend of the Ngo Dinh family, Andre Nguyen Van Chau, recalled the gentleness of Diem's character when he noted that the man never liked being harsh with anyone. Despite the portrait drawn by his critics, Diem was not a violent man, nor even the authoritarian type" (2015, 25).

Diem's ability to cooperate with US policymakers and military leaders was the critical ingredient needed to implement a successful counterinsurgency campaign against communist forces in Vietnam. Against the recommendation of American advisors, Diem believed that a decentralized government would threaten the war effort by bringing in outsiders to the decision-making process. He was very much opposed to delegating duties and had a foundational distrust of outsiders. As Jones (2003) explains, "Diem's stand against decentralization ran deeper than wartime considerations. To delegate more authority to field officers would provide the military with the means for staging another coup attempt" (Jones 2003, 29). American advisors had erroneously assumed that an anti-communist leader with a disciplined disposition would be the ideal partner in the fight against the opposition. They quickly realized that Diem was not amenable to their ideas and very much opposed associations with outsiders in general.

CONSCIENTIOUSNESS

According to Judge et al. (2002), conscientiousness is described as a tendency to be organized, meticulous, detailed, thorough, controlled, and decisive. While Diem felt strongly about his faith and family values, he was conflicted between an attraction to a religious life and responsibilities of leading a country. As Shaw (2015, 37) notes, "Many years later, in a moment of exasperation with Diem's political naiveté during the Buddhist crisis of 1963, his brother Ngo Dinh Nhu shouted that he should have been a monk and not a president."

Jones asserts,

> Diem was not superstitious; he was very rational and pragmatic. However, he had lost all touch with reality and feared a coup more than he feared the Vietcong. His refusal to delegate authority had resulted in administrative bottlenecks that obstructed mobilization of resources and seriously damaged the war effort. His unbending resistance to delegating authority ensured continued frustration in Washington and prolonged instability in Saigon. (2003, 108)

PUBLIC DISSENT

While the Vietnamese people trusted and respected Diem, his government's repressive actions against the Buddhist monks was beginning to affect popular support, thereby offering the opposition a new vulnerability to exploit. One of the primary factors resulting in Diem's eventual downfall was the government's crackdown on a Buddhist religious protest in Hue on 8 May 1963. The protests stemmed from the people's frustrations with politics and repression. They viewed Diem's government as being autocratic in nature. During this protest, which occurred on the birth date of Buddha, nine people were killed and another fourteen injured. The image of a Buddhist monk performing self-immolation, or setting fire to himself as a form of sacrifice, was captured by international press and viewed by millions of people all over the world. Diem had a Buddhist struggle and movement on his hands. This was not merely a dissident religious minority movement. Rather, it created distrust between Diem and the people of Vietnam and marked the beginning of his fall from power.

A declassified Department of State telegram from Saigon to the Department of State dated August 24, 1963, at 4:37 a.m. states,

> Conflicting figures given on casualties at Hue. General Tri states no one killed, 8 Buddhists slightly injured and 2 needed hospitalization, 10 combat police slightly injured and 6 seriously injured, and four ARVN soldiers seriously injured. Some Buddhists killed. Figure of 30 not impossible by any means. In Qui Non there are reports that homes of four Buddhists were broken into and ransacked on morning of August 23 and Buddhists arrested. (*Foreign Relations of the United States*, 1961–1963, Vol. IV, Vietnam, August–December 1963)

The Buddhist protests revealed the glaring extravagance and excessive cruelty of Diem's regime, now led by his well-connected brother and his outspoken wife Madam Ngo Dinh Nhu. A journalist's well-timed photograph of a monk setting himself on fire in the streets of Saigon reached an international audience almost immediately. One photograph captured a fleeting moment in history and the conflict between the Catholic South Vietnam president and the Buddhist monk majority population. Photographer Malcolm Browne recalled the event,

> It was a particularly hot day. The monks suddenly rose and we walked out into the street and began moving in a long column toward the center of town. An old Austin automobile, carrying several monks, led the procession. That struck me as unusual. At that point, the old Austin drove up to the center of this group and out stepped an old man, a very old monk, who I learned later was Thich Quang Due, and two younger monks. He was resting his hands on their arms, and going over to the center of this circle. They put down a tan cushion in the middle of the asphalt. . . . It is a horrible recollection.
>
> The young monk went back to the car and got out a polyethylene jerry can, filled with pink gasoline, which I learned later had been diluted with jet fuel to

> make it burn for a longer time. They poured it over his head and took a few steps back. At that point the elderly monk pulled out a box of matches and struck one of them and dropped it in his lap. And the flames engulfed him immediately. He winced. You could tell from his expression that he was in terrible pain, but he never cried out. He burned for 10 minutes or so, or perhaps a bit more; it seemed like an eternity. The whole intersection smelled of roasted flesh, and the monks and nuns were crying out and screaming. Oh, the screaming and moaning. (Hal Buell 2013, 43–44)

GLOBAL INSTABILITY

Diem's regime in South Vietnam received US assistance against continued North Vietnamese communist efforts. The communist threat was pervasive and seeped into every aspect of US foreign policy. In the case of the Republic of Vietnam, the communist threat was very real, and its presence created violence and wreaked havoc on the Vietnamese people. Ho Chi Minh's communist guerillas had sided with the Soviet Bloc. Consequently, US policy makers feared a "Domino Effect" such that if the Republic of Vietnam fell under communist control, then the rest of the Southeast Asian countries would experience the same fate. As a result, Washington was more than supportive in assisting an anti-communist leader who appeared to be disciplined and restrained in his words and actions. Even at this time, the Republic of Vietnam was still in the early phases of development as an independent country. Moreover, US intervention on the Korean Peninsula in June 1950 in response to North Korea's invasion of South Korea represented a comparable concern.

The events leading up to the assassination of Diem changed the course of history for US foreign policy toward the Southeast Asian nation. It also brought a sense of realism to American policymakers in a way that forced them out of looking at every country through Cold War lenses. Ho Chi Minh's bloody insurgency also forced the US military to reconsider its stance on conventional warfare. Large tanks, massive amounts of weaponry, firepower, and soldiers were discovered to be ineffective against an unconventional enemy that hid in the shadows of the jungle. It was during this time that President John F. Kennedy brought the US Army Special Forces into roles of military advisors and specialists in unconventional warfare. These tactics allowed the Green Berets to train indigenous forces, and it was an effective means of fighting an enemy while leaving a small footprint.

INTERNAL CONFLICT

The internal threat of communist was very real and very violent. As Jones explains,

> Many Vietcong members were married to South Vietnamese women or were brothers or sons of people in the south. Thus, the Vietcong could easily intimidate relatives into collaboration. When Diem took office, the outgoing French controlled only the main towns and highways. During the next two years he had worked to regain the countryside and establish internal security. He had been so successful that the Vietcong resorted to terrorism. (2003, 36)

Vietnamese society and the concept of a nuclear family were centered on Confucian philosophy. In actuality, the Vietcong was known as an agrarian peasant movement.

Notorious for brutal and violent tactics against their own people, Vietcong methods had intensified in brutality since the end of 1960. Until then, the Vietminh had sought to undermine the Diem regime by propaganda and terror. However, the pace of activity had quickened along with its savagery. Millions of party pamphlets propagated lies about the Saigon government, while the Vietcong engaged in abductions, murders, and mutilations. When the Vietcong hit the capital of Phuoc Thanh province in mid-September 1961, they brazenly executed its chief, his assistant, and ten civil servants and inhabitants, including a woman and child. As recently as October 12, 1960, the International Control Commission received reports of 806 deaths and 770 kidnappings. Photographs recorded grisly sights: numerous beheadings of women and children along with government officials and teachers. The Communist Party spread the saying, "kill the land robbers, to encourage the wanton seizure of land from Diem's supporters" (Jones 2003, 103).

REGIME TYPE

In the disputes between China and the Soviet Union, Minh sided with the Soviet Union. Diem's philosophy was Confucian in principle, emphasizing a bureaucratic order that placed him at the top as the Son of Heaven, served by well-educated civilian and military figures known as Mandarins, whose authority extended down to the district and provincial levels. The relationship between the United States and South Vietnam soured once the former included in its requirements that Diem forego his Mandarin principles in order to embrace democratic reforms—something he was not willing to do.

The assassinations of Diem and his brother and advisor Ngo Dinh Nhu on 2 November 1963 changed the trajectory of the war in Vietnam. Not more than three weeks later, President Kennedy was assassinated in Texas. Lyndon B. Johnson assumed the presidency and continued the policy to keep American troops in Vietnam. According to the Assassinations Report issued by the US Senate Select Committee on Intelligence in December of 1975, the initial coup plan failed because "the Generals did not feel ready and did not have sufficient balance of focus" (Assassinations Report 1975). In addition, on September 12, 1975, Senator Frank Church, with President Kennedy's endorsement, introduced a resolution in the Senate denouncing the Diem regime for its "repressive measures" and called for immediate discontinuation of aid to the country (Assassinations Report 1975).

On 1 November 1963, ARVN soldiers took over police headquarters and radio stations in Saigon and began their way to the presidential palace. It was at this time that they confronted Diem and demanded that he give up control of the country. Diem, in turn, called US advisor Henry Cabot Lodge, who told him that there was nothing Washington could do at this time. Diem and Nhu attempted to escape through a secret tunnel but were discovered and taken into custody. Shortly thereafter, both were assassinated in an armored vehicle while being transported to a military headquarters building. The Assassinations Report concluded by stating, "The details of Diem's and Nhu's deaths are not known and none of the informed sources give any indications of direct or indirect involvement of the United States" (The Assassinations Report 1975). According to the Assassinations Report, CIA Director John McCone later stated after a meeting with both the president and Attorney General Robert Kennedy, he believed that President Kennedy agreed with his recommendation to assemble intelligence on the coup plot. The Assassinations Report also stated that an unnamed CIA station chief in Saigon cabled Washington that he had recommended to Ambassador Lodge that the United States "not set ourselves up irrevocably against the assassination plot" (Assassinations Report 1975). When President Kennedy heard about the assassination of Diem and his brother Nhu, it was reported that he demonstrated shock at the news. President Kennedy had not officially approved plans to assassinate Diem. Rather, it had been agreed and assumed by US leaders at the time that the course of action would involve withholding aid to Vietnam and sending more US troops to the country. General Don, one of the coup plotters, alluded to Ambassador Lodge that it was too late, and nothing could stop the coup from happening. This case study showcases how Eisenhower, Dulles, Lodge, Kennedy, and other US policymakers viewed events in the Republic of Vietnam through a Cold War template. It was crucial to keep the country out of communist influence.

CONCLUSION

The overt influences of China and the Soviet Union instilled a sense of urgency over US policymakers' containment policy. The establishment of NATO in April 1949 created the impetus for more collective security in Asia. It was at this time that several economic and military assistance programs were inaugurated.

According to Jones,

> The Diem coup was one of those critical events in the history of US policy that could have altered our commitment. The choices were there: (1) continue to plod along in a limited fashion with Diem—despite his growing unpopularity; (2) encourage or tacitly support the overthrow of Diem, taking the risk that the GVN might accommodate to the VC; and (3) grasp the opportunity—with the obvious risks—of the political instability in South Vietnam to disengage. The United States had two major options: either promote a coup, which it preferred not to do, or enhance the American military presence, which more than a few presidential advisers were willing to do. (2003, 109)

The Republic of Vietnam case study highlights the significance between internal conflict and public dissent. The presence of these two primary factors has a direct influence on the other, and, ultimately on the fate of the leader. Of all the case studies presented in this study, the story of Vietnam is the most puzzling. American diplomats knew very little and appeared to care even less for the culture and traditions of South Vietnam in the 1950s. It was even more surprising to them as well as to US policymakers at the time that nationalism could turn into violence and chaos as illustrated by the Buddhist protest and the rise of terrorism through the Vietcong. Moreover, the divisive nature of South Vietnamese politics, combined with a Catholic leader with Confucius roots, created a pendulum-like situation in Washington that swung from disinterest to investment in preventing a Southeast Asian country from falling toward communism. According to President Dwight Eisenhower's domino theory, it was crucial that communist expansion not occur.

Finally, this case study analyzed the effects of five causal conditions on the fate of leaders to include leadership characteristics, public dissent, global instability, internal conflict, and regime type. In regard to leader characteristics, Diem was not a member of the minority population, he did hold a college degree, and was over the age of forty years old when he assumed power (he was fifty-three years old). Public dissent was not as significant of a factor, given that he enjoyed a tremendous amount of public support from the Vietnamese people. He was viewed as very disciplined in his actions and deeply rooted in his Catholic faith. He assumed power during a tumultuous time when the country was on the verge of gaining independence after enduring colonialism under French rule. With respect to global instability, the threat of communism was strong and heavily influenced foreign policy behavior of the United States. Regarding internal conflict, the country was experiencing newfound independence following French colonialism. As pertains to regime type, the South Vietnam government was an aspiring democracy that was still bound by autocratic rule. The government of the north had communist leanings.

Looking at events during that time, it can be assumed that the assassination of President Kennedy merely three days after the Diem coup may have overshadowed events in Vietnam. Nevertheless, the case study is one that is rich with history, and one cannot help but want to seek more ways to learn about how this small Southeast Asian country changed the US military's doctrine on conventional warfare.

Table 8.1.

Variable/ Membership	*Fully In*	*Mostly In*	*More In Than Out*	*More Out Than In*	*Mostly Out*	*Fully Out*
Leader Characteristics				X		
Public Dissent	X					
Global Instability				X		
Internal Conflict	X					
Regime Type				X		

Chapter Nine

Chile, 1970–1973

This chapter explores covert action events planned and carried out by the United States in Chile from 1970–1973. It starts with a brief overview of the political environment during that time and identifies and explains the leader characteristics of Chilean president Salvador Allende and the effects of his leadership and policies on Chilean society. After discussing the overall aftermath of the covert action operation and explaining how American forces used that action to influence the Chilean government, the study will conduct a detailed discussion of the degree of membership this case demonstrates for each factor presented in this study's QCA model. Finally, this case study concludes with a summary of the findings and an application of the findings into the degree of membership matrix.

The September 1973 coup in Chile, facilitated by the United States government, created an idea that a superpower could overthrow a democracy, especially if there was any semblance of a communist threat. Chile was developed by the United States as a sophisticated nation with strong political institutions, which established a solid reputation as a democratic government. As Hove (2007, 626) notes in *The Arbenz Factor: Salvador Allende, US-Chilean Relations, and the 1954 US Intervention in Guatemala*, "When Dwight D. Eisenhower entered the White House in January 1953, he approached US-Latin American relations with a sense of urgency."

BACKGROUND

Characterized by President Harry S. Truman as the "greatest functioning democracy" in South America, Chile was elevated to this status as a result of its commitment to freedom of press, speech, and religion. From its independence in 1818 to the coup of September 1973, Chile had only experienced four brief interruptions in its democratic tradition (declassified memorandum, Senate Select Committee on Intelligence Activities, Gerald R. Ford Library). In comparison to Guatemala and Congo case studies, Chile boasted a literacy rate of an impressive 91 percent—over a majority of its population. Also interesting to note that Chilean ethnicity was based largely on indigenous American and European roots. "Faced with threats from the Right and the Left, US officials navigated a middle, anti-totalitarian course in order to preserve democracy and political stability in Chile" (Hove 2009, 27).

LEADER CHARACTERISTICS

Salvador Allende was born in 1908 into a wealthy family and entered into the political realm while he was a young medical student at the University of Chile in 1926. It was also during this time that Allende cultivated a close friendship with Cuba's Fidel Castro. It is interesting to note that Allende's daughter, Beatriz, married a Cuban diplomat and close confidant of Castro's, thereby creating an even more intimate merger between the two families.

Flynn notes,

> Although Allende trained as a physician, he thought that he would be better able to serve the people through direct political action. In 1933, he helped found the Socialist Party of Chile. The Socialist Party of Chile adhered to many basic socialist ideals including the belief that capitalist exploitation of the land must be stopped and that the state of private property must be discontinued and transformed into a system of collective ownership. (2014, 281–282)

NEUROTICISM

Neuroticism is a term used to describe an individual who is highly emotional, highly prone to anxiety, depression, violence, and anger. As a young man, Allende had witnessed the Chilean government's control over its people. Despite his privileged upbringing, a fact he was often embarrassed to admit to his followers, Allende felt a close association to the needs of the Chilean people, and he devoted his life to fighting for justice and equality.

EXTRAVERSION

Extraversion is described as personality types that seek excitement and experiences beyond their comfort zone. Essentially, it is the quality of being confident and outgoing.

As Hove describes, "Allende himself consciously used the rhetoric of Marxism within his party but nonetheless tried to maintain the external image of a democrat. His personal habits did little to discourage this impression. Allende was known for his conspicuous consumption, his lavish lifestyle, and his personal eccentricities" (2007, 27).

Former US defense attaché in Santiago, Colonel Paul Wimert adds,

> This beautiful yacht came gliding across the water to hoop up to the wharf. There was this gentleman in a white hat and a white cravat with a blue admiral's coat and white duck pants and shoes. He looked [like] one of the old wealthy people—the Vanderbilts of America—I couldn't believe it was Allende . . . he was beautifully groomed, and of course his sailing ship was one of the nicest ones there. It didn't sit right that he was supposed to be communist or socialist. (Wimert 1999)

In a memo to President Lyndon B. Johnson, McGeorge Bundy notes, "The problem we face is that a very popular and attractive candidate, named Allende, who is thrown in his lot with the Communists, has more than a fighting chance to win" (McGeorge Bundy, "Presidential Election in Chile," memorandum to Lyndon B. Johnson, May 13, 1964, NSF, Country File, vol. 1, 1/64–f8/64, Memos to the President, McGeorge Bundy, vol. 4, LBJPA, https://nsarchive2.gwu.edu/coldwar/interviews/episode-18/wimert1.html).

According to the KGB archives, "Allende's characteristics traits were arrogance, vanity, desire for glorification and a longing to be in the spotlight at any price. He was easily influenced by strong and more determined personalities. He was also described as something of a womanizer" (Andrew and Mitrokhine 2006).

OPENNESS TO EXPERIENCE

This type of personality describes individuals who are bold and adventurous who dislike routine, prefer abstract thinking, and have a wide range of interests. The opposite of this personality trait is individuals who are traditional, non-artistic, and non-analytical.

Hove notes that,

> The KGB issued the appropriate code name LEADER to Allende, though the Chilean was not officially recruited as an agent nor did he ever take KGB payments. Systematic contact between Allende and the KGB was established in 1961, when relations were restored and the KGB set up a residency in a new trade mission. To the KGB, Allende had stated his willingness to cooperate on a confidential basis and provide any necessary assistance, since he considered himself a friend of the Soviet Union. He willingly shared political information. (2007, 30)

AGREEABLENESS

This personality trait describes individuals who are compassionate, trustworthy, lenient, and softhearted. In contrast, it can also describe individuals who are hostile, cynical, manipulative, vengeful, and ruthless.

CONSCIENTIOUSNESS

The personality trait of conscientiousness describes individuals who are able to manage their impulsive behavior through persistent motivation. These individuals also prize perfectionism in their quest for success and status.

Flynn goes even further to explain:

> Allende ran unsuccessfully for president in 1952, 1958, and 1964. Each time he gained more support and as his political strength increased he became well known on an international level. Unlike other leaders who became jaded or simply give up, Allende became more determined to bring democracy to Chile.
>
> In his September 5, 1970, victory speech to the citizens of Santiago, Allende expressed a deep appreciation for the support of the people. Allende referred to himself as the "companero president" and Chile as "their country." He also commented that a revolution "does not mean to destroy but to build." Allende's speech was eloquent and simple; dramatic and calm; practical and hopeful. (2014, 282)

PUBLIC DISSENT

During US covert operation planning against the newly democratically elected Allende government, Chile began to experience a deterioration of its social and economic infrastructure as a result of ineffective governance. This was due primarily to the infighting that took place among Marxist parties of the Unidad Popular coalition (UP). Illegal land seizures and unsuccessful management of the government created a chaotic environment that proved a fertile breeding ground for public dissent. Paranoia and distrust were common as a result of the recent kidnapping and assassination of General René Schneider Chereau, commander-in-chief of the Chilean Army during the 1970 presidential election.

In terms of economic stability, a brief growth spurt of foreign reserves was used to boost the salaries of public employees whose ranks had swelled shortly after the election. The Chilean people grew increasingly dissatisfied with the inefficiency of government management executed by the UP coalition. The first major protest occurred on 1 December 1971 when thirty thousand women protested in the streets. At that time, several UP coalition members were removed from their government positions and a drop in copper prices and agricultural production began simultaneously. Moreover, it appeared that the country was precariously close to an economic collapse.

Gustafson informs,

> Beginning in mid-1972 a series of violent and protracted strikes slowly destabilized the UP government. The first of these began on August 21 and was led by the nation's independent shopkeepers. Strikes and street violence followed and continued until police closed off downtown Santiago after a two-day running street fight between youth arms of the rival political parties MIR and Patria y Libertad. (2007, 148)

GLOBAL INSTABILITY

In 1964, US interests in Chile were very clear—to prevent the ascension of a Marxist leader who could cause the United States economic impairment and allow unwanted Soviet influence in the Americas. That Marxist leader was Allende, who rose to power during a time of political and economic chaos in Chile. Additionally, the Americans viewed him suspiciously for his Marxist leanings to other Marxist leaders in the Americas.

In a Memorandum for Dr. Henry Kissinger from the CIA, dated 13 September 1973,

> This agency did not conduct operations in support of either of the two democratic candidates who opposed Salvador Allende in the 1970 presidential election. Our role in the election was limited to an effort to denigrate Allende and his Popular Unity coalition during the campaign. After the March 1973 elections, it became increasingly apparent that three years of political polarization had strained the fabric of Chilean society to the breaking point. (1973)

INTERNAL CONFLICT

Beginning in 1961, the administration of John F. Kennedy formed a committee of top-level officials from the State Department, the White House, and the CIA. In addition, there were perhaps one hundred intelligence operatives working within Chile. In the years leading up to the 1964 election, these intelligence agents began establishing relationships and providing funds to Chilean political parties that opposed Allende.

As the election neared in 1964, the United States decided that the best candidate was the centrist Eduardo Frei of the Christian Democratic Party. To aid Frei, the CIA covered approximately half of his campaign costs, spending up to an estimated $20 million in Chile. The CIA also launched a massive anti-communist campaign to alarm the general population of what communism would mean in Chilean life. Furthermore, US intelligence operatives subsidized the anti-Allende media and provided funds for grassroots organizations that promoted reform rather than left-wing revolution. US intelligence efforts in Chile succeeded, as Frei won 56 percent of the votes compared to Allende's 39 percent.

The United States continued its intelligence efforts in Chile in the years after the 1964 election. In the 1965 and 1969 Chilean congressional elections, America backed anti-Allende candidates. Despite continued US efforts, in the 1970 presidential election, Allende won with a plurality of the votes. Due to the fact that Allende did not win a majority, the Chilean constitution stipulated that the Congress would choose between Allende and runner-up Jorge Alessandri. The United States continued to prevent Allende from taking power.

However, the commander-in-chief of the Chilean Army, René Schneider, insisted on respecting the constitutional process. Then, in a kidnapping attempt, Schneider was killed. However, instead of allowing the coup to go forward, Schneider's murder discredited the right-wing political parties and rallied the Chilean military around the country's constitution, allowing Allende to become president.

After failing to prevent Allende from taking office, US intelligence sought to destabilize Chile by targeting the country's economy. The United States government provided virtually no economic aid to the Chilean government and pressured the United States Export-Import Bank, the Inter-American Development Bank, and the World Bank to withhold aid. Furthermore, Chilean businesses had difficulty acquiring goods from US suppliers. Soon the country became politically polarized between pro- and anti-Allende camps.

The United States also increased its military aid to Chilean dissidents and expanded the training of Chilean military personnel in the United States and

Panama. Furthermore, the CIA encouraged and aided labor strikes in the country. For example, in October 1972, US intelligence supported a strike by private truck owners, which disrupted the flow of many goods. Later, bus companies went on strike, as did many professional workers. The CIA continued to influence the media and sought the support of the military against Allende. Even in Washington, DC, US intelligence bugged the Chilean Embassy and stole documents. Despite these efforts, in the March 1973 congressional elections, Allende's coalition won 44 percent of the votes, a greater percentage than it had won in 1970.

In September 1973, the Chilean military acted, staging a coup that overthrew Allende. Gen. Augusto Pinochet led the military coup that caused the death of Allende and marked the beginning of a brutal dictatorship that lasted until 1989. Pinochet abolished political parties, dissolved Congress, banned union activities, and suppressed civil liberties. The Pinochet regime was also known for its human rights abuses, as thousands of Chileans were tortured or killed.

According to Meade,

> Beginning in 1974, prisoners were brought to the nearby airstrip and from there they were loaded onto trucks and brought the short distance to Villa Grimaldi. Those brought via the airfield were combined with the people who were trucked in from neighborhoods, workplaces, schools, and universities throughout the city. The trucks pulled up to a wooden gate, which remains today, complete with the sliding door that had enabled the guards to check the identity of the driver and his cargo. The prisoners, all with arms tied behind their backs and blindfolded, were unloaded in much the same way that livestock had been delivered to the villa when it was a country estate. According to testimonies from ex-prisoners, six agents "greeted" the arrivals with clubs. They were then taken to the "grill," a converted metal-frame bed, where they were subjected to one to one-and-a-half hours of electric shock torture, not quite enough to kill them. Those who did pass out or suffered heart attacks were revived by torturers (trained by military physicians) who either threw cold water on the victims or performed CPR to keep them alive. Variations on the standard shock torture included turning the "rack" of the bed upside down or conducting the torture in front of the prisoner's friends and family members who were also being held. (2012, 129)

REGIME TYPE

The CIA's covert action strategy relied heavily on propaganda in its attempt to influence various sectors of Chilean society. One avenue was to spread the pro-Christian Democratic message throughout the urban cities within the slums as well as outside the urban environment to include villages and towns outside of the main city. The CIA also heavily endorsed women's groups that advocated an anti-communist message. They also supported university student protests as well as women's group marches.

According to the Staff Report of the Select Committee to Study Governmental Operations with respect to Intelligence Activities: United States Senate,

> Covert United States involvement in Chile in the decade between 1963 and 1973 was extensive and continuous. The Central Intelligence Agency spent three million dollars in an effort to influence the outcome of the 1964 Chilean presidential elections. The range of clandestine activities undertaken by the CIA includes covert action, clandestine intelligence collection, liaison with local police and intelligence services, and counterintelligence. (Staff Report of the Select Committee to Study Governmental Operations with Respect to Intelligence Activities: United States Senate, published by the US Government Printing Office, Washington: 1975, 148, United States. Congress. Senate. Select Committee to Study Governmental Operations with Respect to Intelligence Activities: Alleged assassination plots involving foreign leaders: an interim report of the Select Committee to Study Governmental Operations with Respect to Intelligence Activities, United States Senate: together with additional, supplemental and separate views. (Washington: US Govt. Print. Off., 1975)

When the coup attempt failed and Allende was inaugurated president, the CIA began funding opposition sectors in Chile.

The precedence set forth by US leaders in regard to Chile was that no amount of money was too great to overthrow a communist regime. After the failure of the coup and Allende's eventual inauguration as the new president, the US still continued to pour millions of dollars into subverting the political system in Chile. Chile had initially begun as a model for democracy and was viewed by the US as a platform for which to launch new development programs. Ultimately, its tendency to lean toward communism was the critical factor that changed the course of history. It took Allende four times as a presidential candidate to finally win the presidency. A self-proclaimed Marxist, Allende was able to persuade the people's support through his grassroots approach to the country's development.

CONCLUSION

One of the interesting associations to note in the Chilean case study is the link between leader characteristics and global instability. The United States viewed Allende with suspicion, and the CIA devoted much of its time to attempting to prevent his ascension to power, and when that tactic failed, to assassinating him. The external influence from the United States had a direct influence on the leader's demise. Allende's flamboyant leadership style, combined with his inability to manage his country's economic and political systems effectively, provided the CIA with ample opportunities to dispose of him. Allende's election was viewed as a defeat and direct threat to Chile's long-standing democracy. It was, to that point, a model democracy.

This case study analyzed the effects of five causal conditions on the fate of leaders, to include leadership characteristics, public dissent, global instability, internal conflict, and regime type. With respect to leader characteristics, Allende was a member of the minority population, he had a medical degree, and he was under the age of forty years old when he assumed power. Public dissent was a significant factor as the country was thrown into chaos and economic upheaval as a result of political infighting between Marxist groups and governmental mismanagement of the country. He assumed power during a time when Chile was held in high regard by the United States. As pertains to global instability, the threat of communism was strong and heavily influenced foreign policy behavior of the United States. Regarding internal conflict, the country was not in a state of crisis at the time of the covert action event. The economy was strong and many of the Chilean people were well off economically. With respect to regime type, Chile's long-term democracy was viewed in high regard by the United States. It was not until Allende's election that the United States expressed concerns about the fragility of Chile's role as a model democracy in the Americas.

Table 9.1.

Variable/ Membership	*Fully In*	*Mostly In*	*More In Than Out*	*More Out Than In*	*Mostly Out*	*Fully Out*
Leader Characteristics	X					
Public Dissent		X				
Global Instability	X					
Internal Conflict		X				
Regime Type			X			

Chapter Ten

Political Systems

Regime Change and Leader Survivability

A dictator's life is ephemeral. Violent coups, torture, execution, assassination by enemies or even close friends or family are occupational hazards. Escribà-Folch notes, "As for regular or nonviolent exits, restrictions on civil rights can thwart the ability of some actors (especially the opposition) to organize, coordinate, and mobilize in the short-term" (2013, 546). In *Fighting for Survival: The Fate of Leaders and the Duration of War*, Goemans offers, "Dictators stay in power largely thanks to their effective repressive apparatus. Once they lose power, however, dictators are highly likely to suffer severe additional punishment, such as death, imprisonment, or exile" (2000, 558). This project expanded on that study by analyzing specific leaders and their survival during specific covert action events.

In his seminal work, *Political Order in Changing Societies*, the late Samuel Huntington suggests that "the most important political distinction among countries concerns not their form of government but their degree of government" (1968, 1). He contends that violence in some societies "was in large part the product of rapid social change and the rapid mobilization of new groups into politics coupled with the slow development of political institutions" (1968, 4). In many of the case studies analyzed in this research, leader survival was directly affected by the individual's desire to push toward modernity, thereby resulting in their inevitable demise. In essence, modernity breeds stability but modernization also breeds instability.

During European colonialism, authoritarian governments were created as a means to control territories that were not suitable for permanent settlement. During this process of colonialism, European states such as France, Portugal, Spain, the Netherlands, and United Kingdom, set up colonial regimes that served the economic, social, and political interests based on the interest of their own domestic interests, while ignoring the economic and political development of the exploited countries.

Such was the case in Africa, where European colonists were authoritarian and more concerned with effective administration than political representation (Bratton 2009). Either they ruled directly through coercion or they devised indirect control by piggybacking Western laws on existing systems of traditional authority (Bratton 2009). Bratton also states that only in belated reaction to mass African nationalism did the departing colonial authorities try to install institutions of (British) parliamentary or (French) presidential democracy (Bratton 2009, 341). European delegates relied on elite leaders of the colonized country to spread the message of colonialism. In turn, these elite leaders became economically linked to the West. In this respect, newly independent African nations briefly partook in the second global wave of democratization that accompanied the break-up of European empires in the middle twentieth century (Huntington 1991). The result was no political modernization of any kind. Levitsky and Way describe the centrality of informal institutions as one characteristic of competitive authoritarianism, which emerged in Africa's post-colonialism period (2010).

Historically, regime change operations have occurred in waves with the great power or powers conducting such endeavors. This research provided fertile ground for the exploration of international relations theories as a template to analyze covert action case studies. During the Cold War, the international system was bipolar, and the two superpowers of the United States and Soviet Union held the most influence. There was no question about who held the power. Classical realists such as Hans Morgenthau and Reinhold Niebuhr believe that states are similar to human beings in that each has an innate desire to control others. They argue that this self-serving desire for power is the cause of wars. In *Politics among Nations*, Morgenthau (1948) stresses the value of a multi-polar balance of power system and notes that realism is fundamentally concerned with power and not morality or material interests.

NEOREALISM

In *International Relations: One World, Many Theories*, Stephen Walt notes that "the neorealist theory advanced by Kenneth Waltz ignored human nature and focused on the effects of the international system" (1998, 2). He adds, "For Waltz, the international system consisted of a number of great powers, each seeking to survive. Because the system is anarchic, each state has to survive on its own. Waltz argued that this condition would lead weaker states to balance against, rather than bandwagon with, more powerful rivals" (1998,

2). International relations theories can be compared to lenses on a camera: each theory offers a different way of looking at a social phenomenon from a series of different angles. The lenses of realism might be seen in the starkness and black and white while the complexities of neoclassical realism may be seen in a wash of gray tones.

NEOCLASSICAL REALISM

The framework for this project was based on the international relations of neoclassical realism. As Taliaferro states in *State Building for Future Wars: Neoclassical Realism and the Resource-Extractive State*, "Neoclassical realism merges neorealism's emphasis on systemic constraints—constrictions caused by the incessant quest of states for 'survival' in the anarchical global realm—with classical realism's focus on the dependence of statesmen in acquiring domestic material power resources for regulating the magnitude and ambition of their foreign and defense policy aims" (2006, 480–481). He adds, "Neoclassical realism argues that over the long run, the relative amount of material distributions and the existing context of the international system will engineer states to maximize their global influence, strength, and security" (2006, 480–481). Furthermore, a central tenet of neoclassical realism is that "states use the tools at their disposal to gain control over their environment" (Rose 1998, 157). According to neoclassical realists, "decision-makers' beliefs strongly affect the relationship between relative power and foreign policy" (Rose 1998, 157). Moreover, he describes the *Four Theories of Foreign Policy* as

1. Innenpolitik describes foreign policy as a direct outgrowth of domestic politics and factors such as ideology, culture, and economics shape states' foreign policies (Rose 1998, 149).
2. Offensive realists—or aggressive realists—describes a classical realism, Hobbesian world in which states seek to maximize what little security they have and, as a result, foreign policy consists of "nervous states jockeying for position within this anarchic framework" (Rose 1998, 149).
3. Defensive realists address external threats through subtle modifications of the power balance. As a result, "foreign policy consists of (largely peaceable) reactions to systemic factors" (Rose 1998, 149).
4. Neoclassical realists reject the security-only thesis; instead, "states attempt to use their power to direct the international system towards their own goals and preferences" (Rose 1998, 149).

All covert action employed similar tactics: the US-supported anti-Soviet dissident movements in order to overthrow the central government and establish an independent regime, one more favorably disposed to, if not formally aligned with Washington. In his book, *The Working Class in the Hungarian Revolution of 1956*, Lomax writes, "The Hungarian Revolution of 1956 has been portrayed as a revolution primarily for freedom and truth—it arose to restore the individual and intellectual freedoms which had been abolished by the totalitarian communist state" (2007, 27). Granville (2005) adds, "Flush with the success of the covert operations in Iran (1953) and Guatemala (1954) to remove leaders of suspected leftist or pro-Soviet tendencies, Eisenhower and other US policymakers in the 1950s appeared blind to the existence of scrupulous, reform-minded Communists" (2005, 839). She concludes that "this prejudice led them to discredit national Communists such as Imre Nagy of Hungary and Ho Chi Minh of Vietnam and to prefer non-Communists no matter how unsuitable their personalities for national leadership" (2005, 839).

CONTRIBUTION OF THE STUDY TO EXISTING BODY OF LITERATURE

This research contributes to the existing academic literature by analyzing whether personality traits and leadership styles affect an individual's survival following US covert action. The existing literature on the consequences of regime change focuses on successful overt operations and analyzes their effects on numerous variables, including civil war, democratization, and state building. Other studies focus on leader longevity and the factors that attribute to a long or brief tenure in office. Additional scholarly books discussed leadership structure in an organization (hierarchy or decentralized) and whether this has an impact on survival of the organization. Moreover, this project complements this literature by analyzing the effects of covert regime change outside the context of war and analyzing its effects on leaders' survival. The unique aspect of this research is that it delves into the political and sociocultural contexts of the targeted country at the time of the event. Each case study offered the researcher a rare opportunity to become fully immersed in the personal and private lives of each leader. The personal aspects of a leader are documented and meticulously researched in order to uncover the driving force behind a leader's future development, personal ambitions, private vices, and insatiable appetites for power, control, knowledge, and compassion or, in some cases, cold-blooded brutality.

This research addresses the following research question: "To what extent, if any, can leader survival be predicted based on analyses of US-sanctioned covert action events during the Cold War?" From a review of the relevant literature on covert action, international relations, and democracies, the following model was produced: Survival (s) = Leader Characteristics (lc) + Public Dissent (Pd) + Global Instability (gi) + Internal Conflict (IC) + Regime Type (RT) or s= lc+Pd+gi+IC+RT. In addition to the hypothesis that, collectively, these factors determine the outcome of leader survival, it was further proposed that leadership characteristics and regime type would have the most influence on an outcome.

THE WESTMINSTER MODEL OF DEMOCRACY

In *Patterns of Democracy*, Arendt Lijphart (1999) calls readers' attention to two competing styles of democracy: the Westminster model of democracy (or *majoritarian* because it is a model represented by the majority), named after Great Britain's system, and the consensus model of democracy, which Lijphart refers to as the "kinder, gentler" democracy. According to Lijphart (1999), the Westminster model is lacking in some ways because it excludes almost half a given population from the governmental process. The "majoritarian interpretation of the basic definition of democracy is that it means government by the majority of the people in that majorities should govern and that minorities should oppose" (Lijphart 1999, 31). He adds, "This view is challenged by the consensus model of democracy" (Lijphart 1999, 31). Lijphart points out that "in too many societies such as those with deep ethnic, linguistic, religious, or ideological divisions, there is little chance of integration between minority and majority interests in society" (1999, 31).

THE CONSENSUS MODEL OF DEMOCRACY

Lijphart is well-known for his support of the consensus approach, which, in his view, offers more compromise and significant minority rights. He adds that, contrary to popular belief, Westminster democracy does not outperform consensus democracy. According to Lijphart, this type of democracy demonstrates "kinder, gentler qualities because they are more likely to be welfare states, they have a better record of protecting the environment, few people are incarcerated and there is less use of the death penalty, and they tend to be more generous with economic assistance to nations in need" (Lijphart 1999, 276).

PRESIDENTIALISM VS. PARLIAMENTARISM

Parliamentary and presidential forms of government are different in that a parliamentary system is comprised of a separate head of state and head of government, whereas one official fulfills both functions in the presidential form of government (Lijphart 1999, 117). Presidents are elected in national elections and prime ministers are selected in local government. Another difference is that presidential terms are fixed, while parliamentary are flexible. In addition, cabinet members are not members of the legislature in a presidential system as opposed to the parliamentary system where cabinet members must be in the legislature. Moreover, the president can only be removed through impeachment in the presidential system, whereas a parliamentary government must maintain the confidence of the legislature or face dissolution and a new election.

In *Some Social Requisites of Democracy*, Seymour Lipset (1959) seeks to find the conditions of a "stable" democracy, which he explains is necessary for the survival of the regime. In *Democracy and Development*, Przeworski (2000) also studies regime stability and its role in the success of democracies and dictatorships. Coinciding with regime stability, Frey and Brooks both argue that certain freedoms, one of which is economic freedom, are positively correlated with democracy (Frey and Stutzer 2000, 421). The level of the effectiveness of a government is usually considered in the many variations of defining and measuring democracy, but it is not necessarily a condition exclusive to democracy. Dictatorships can also be effective forms of government.

Chapter Eleven

Conclusion

Mossadegh was described as highly emotional, an aristocrat, and a gifted orator. He shared similar characteristics with Indonesia's Sukarno, the Democratic Republic of the Congo's Lumumba, and Cuba's Castro who exuded self-confidence and were well-known as being gifted communicators. Mossadegh, Sukarno, and Castro were physical communicators who tamed the crowd through visible demonstrations of emotions. Lumumba, on the other hand, was more reserved but his speeches were deeply intellectual in content. These leaders had the ability to relate to people on many different levels. It is interesting to note that Lumumba and Sukarno had left their rural homes for life in the big city. Both were greatly influenced by their cultural experiences. Guatemala's Arbenz was known for having an unsteady personality and lack of personal discipline and self-control. He attempted to commit suicide and had multiple bouts with mental depression. The Dominican Republic's Allende was described as a chameleonic person: vain, quick-tempered, easily offended, socially ambitious, charming, and womanizing. Trujillo came from a modest, barely literate rural family and was not considered well-educated. He demonstrated the fascination with personal appearance, clothing, and military decorations that suited him for a military career. Diem was often described as cool and aloof by American advisors. He stood firmly by his Confucian faith and displayed stoicism to the public, even while experiencing tremendous hardship, believing that the traditional family and cultural ways of the Vietnamese should be protected from new, modernized changes. Castro lived austerely, in almost Spartan conditions: there is no need for luxury; his furniture is sober; his food is frugal, healthy, and macrobiotic. His are the habits of a solder-monk. Described as having a thunderous voice, he would often pause in mid-speech, raise his index finger, and point in an effort to tame the crowd (Ramonet 2007, 13). He believed passionately that any change that would come to Cuba would be determined and implemented from within the country. Despite international pressure imposed by commercial, economic, and financial embargoes led by the United States in the early 1960s, Castro remained stalwart in his vision for the country.

Covert action operations in Iraq and Afghanistan will provide fertile ground for future research as records are declassified and the passage of time allows for a more settled perspective on the cases. Additional research could also be conducted on leader characteristics on predicting social movements, coup-proofing, violent uprisings, among others. If the results continue to confirm the findings of this study, the US military and civilian intelligence organizations may consider adopting the factors of this model as part of their covert action planning process.

Another avenue for research lies in a more granular examination of the individual cases examined in this research. Personality profiles such as Patrice Lumumba are rich in detail and there is so much more to explore. The same can be said for the other seven case studies. Each case study offered a glimpse into the personalities and personal lives of the leaders who lived during that time. The most rewarding part of the research process was having the opportunity to enter into each case in order to truly understand the leader's psychological, emotional, and physical conditions. This study can be expanded to include personality profiles on each leader, a comparative study of two leaders with similar or dissimilar outcomes, the factors leading to leader assassination, and the strategic perspective of the impact of leader assassination on a foreign policy.

Moreover, this study highlighted the fascinating correlation between leader survival and US foreign policy. In many cases such as Indonesia, Chile, and the Republic of Vietnam, US senior leaders enjoyed close and personal relationships with the leaders. The most important implication for this research is as a planning tool for planners and military and political leaders faced with conducting covert action events with limited time and resources. The decision to initiate a covert action event should be informed by the presence or absence of the factors or conditions presented in this model. If they are not present or cannot be created, the creation of a covert action program should be rejected. If they are present, planners can use this model as a set of guiding principles when deciding on the organization of a program and as a selection criterion to choose which areas provide a more fertile ground to start and expand the program.

Finally, this book illustrates that political leaders must make the hard decision to provide early support to understanding leader characteristics and cultivating relationships based on these characteristics. Specifically, senior leaders in the civilian and military spheres of influence may also benefit from the research conducted in this study. This research has contributed to the understanding of an understudied element of covert action research, and results of this research confirm the assertion presented in the introduction that the factors contained in this model should be included in covert action planning.

Utilizing the model presented in this study, a leader may be able to make a more educated decision about the use of covert action which may, in turn, save the lives of many. Additionally, the president of the United States has the ultimate authority in the approval of covert action. Having the option to choose between war and diplomacy, the president as well as senior civilian and military leaders must understand how policy shapes the choice between covert action and traditional military engagements. The traditional use of military force is actualized through large, conventional forces requiring complex logistical planning and movement of troops and equipment, in addition to the basing requirements for each operation. These types of conventional wars also serve as symbols of US military power.

While covert action requires meticulous planning, it also offers a sophisticated option that is as influential as large conventional force. Additionally, the 2018 National Defense Strategy clearly states objectives on how to develop a lethal, agile force that is prepared to fight wars in multiple domains. Moreover, it states that the US military should be "strategically predictable" but also "operationally unpredictable" (*National Defense Strategy* 2018, 5).

Finally, regarding the intelligence tradecraft, the CIA should be the critical node for the subtle and sophisticated implementation of covert influence operations while unconventional military forces execute the tactical side. The 2017 National Security Strategy supported the model stating, "To prevail, we must integrate all elements of America's national power—political, economic, and military. . . . The United States must develop new concepts and capabilities to protect our homeland, advance our prosperity, and preserve peace." It is also bolstered by the 2018 National Defense Strategy, which clearly outlines the current and future state of warfare: "Deterring or defeating long-term strategic competitors is a fundamentally different challenge. . . . [O]operations must introduce unpredictability to adversary decision makers. . . . [W]e will challenge competitors by maneuvering them into unfavorable positions, frustrating their efforts, precluding their options while expanding their own, and forcing them to confront conflict under adverse conditions" (*National Security Strategy* 2017).

According to Stretton (1969, 158), "Cross-case analysis examines the commonalities and difference in the events, activities, and processes of case studies. Additionally, it is conducted in order to "extend the investigator's experience beyond the single case by provoking her or his imagination, prompting new questions, seeking new dimensions, measuring alternatives, creating models, and constructing ideal types and utopias." Ragin (1997) adds, "Another purpose of engaging in cross-case analysis is to enhance researchers' capacity to understand 'how' relationships may exist among discrete cases, accumulate knowledge on the original case, refine or develop concepts."

The final analysis conducted was the fuzzy-set truth table analysis. The function of this analysis is to "determine the connections between combinations of causal conditions and an outcome" (Ragin 2008, 109). The standard analysis provides three sets of solutions: complex, parsimonious, and intermediate. The complex case provides a solution that avoids counterfactual cases. The parsimonious case provides the simplest case with the fewest casual conditions through the use of counterfactual cases. The intermediate case allows for inclusion of the most reasonable counterfactual cases. The strength of this analytical method is it determines multiple pathways to a similar outcome.

The parsimonious solution includes two causal pathways: leader characteristics and internal conflict. The overall solution was calculated to have coverage of 0.998175 and a consistency of 0.868254. In this solution, each term is measured by their degree membership in each possible solution for the outcome, and the proportion of each factor contributes to the explanatory value in each potential solution. The importance of this finding to policymakers is that understanding a leader's personality profile is an important consideration in designing a covert action strategy.

The first is leader survival, public dissent, and global instability. The second is internal conflict and regime type. These pathways provide an overall consistency of 0.875614 and an overall coverage of 0.976277. These two pathways are instructive in that they indicate two realistic sets of conditions that can produce a successful outcome. In cases where leader characteristics are strong, regime type are not clearly defined, and where internal conflict is strong, the more advantageous is the opportunity to execute a successful covert action event. Public dissent is also another factor that influences the probability for success in creating a permissive environment. This research contributes to the study of covert action literature in three ways. First, it analyzes political and socio-cultural conditions that existed in the countries. Previous works have focused on these conditions from a US perspective. Second, contrary to most Cold War era covert action studies, this study focuses on leader personality profiles within the context of the environment at that time. Finally, this research focuses on factors influencing leader survival, whereas previous research focused on the CIA's use of covert action, foreign policy from US presidents' perspective, and, more often than not, the failures of these events and their exploitation as depicted by popular culture and media at the time. Covert action will continue to exist, and US presidents will need to know how to execute it smartly. This research also shows the limits of covert action, the nuances of the intelligence tradecraft, the impact of US foreign policy on developing countries, and the long-lasting effects of those decisions. Additional research may include a more in-depth look at the correlation between leader characteristics and other covert action events in history. How do these leaders compare to others? Does geography make a difference? What other factors are influential in determining the course of events? More detail will be revealed once intelligence derived from these events become declassified.

This study proposed a model for utilizing covert action options while remaining cognizant of leader survival. The model correctly hypothesized that the most important factor influencing the success of covert action was the relationship between the leader and the entity executing the covert action event. The model also shows that a balanced adherence to the factors of the model correlates to a successful outcome.

The limited nature of this research requires that additional investigation be conducted, as three of the failed case studies were dropped because of a lack of evidence or a failure to fit the case study selection criteria. More research needs to be conducted on failed cases to confirm the findings of the two failed case studies examined in this study. Specifically, it is important to confirm that a general lack of adherence to the factors of this model consistently produce failure. Additional research could expand the data set or focus on a new set of cases as data from more covert action events becomes available. Future research should consider the impact of technology on the chain of events. For example, would leaders' fate be influenced by the introduction of advanced technology in the form of internet, artificial intelligence, smart phone, and other technologies? It would also be interesting to note the significance of natural resources (oil, minerals, etc.) within the case study country and how it influenced the socio-economic, security, and political environment.

Different types of covert action include covert political action that intends to influence the intentions of a foreign power without disclosing US involvement or intentions; covert media action used to influence public opinion or perceptions through the use of newspapers, magazine articles, radio, television, and the internet; and covert economic action, also called economic warfare, used to affect the economy of a target country, particularly its ability to wage war or sponsor an insurgency in another country. Covert action is intended to achieve national security objectives in a foreign country without disclosing the identity or full intentions of the sponsor; finally, it is used to influence the intentions of a foreign power in the target country producing overt results.

This research project sought to develop a model for covert action decisions that result in leader survival. Additionally, this research uses neoclassical realism theory as the theoretical framework through which to evaluate the relevant covert action literature. The model proposed in this book presents variables believed to influence the potential for leader survival after a covert action event. Specifically, this research takes into consideration the survival of the leaders replaced and their respective replacements as well as details about their personal lives. Despite the large collections of Cold War era covert action literature, many of the works are focused on the success or failure of the events, the role of the CIA in facilitating the events, as well as the relevant literature on the intelligence tradecraft used during these events. Not much, if any, of the literature is focused on covert action from the perspective of the leader who experienced it. The unique aspect of this research is that it delves into the political and socio-cultural contexts of the targeted country at the time of the event. Each case study offered the researcher a rare opportunity to become fully immersed in the personal and private lives of each leader. The personal aspects of a leader are documented and meticulously researched in order to uncover the driving force behind a leader's future development, personal ambitions, private vices, and insatiable appetites for power, control, knowledge, and compassion or, in some cases, cold-blooded brutality.

Moreover, this book addresses the following research question: "To what extent, if any, can leader survival be predicted based on analyses of US-sanctioned covert action events during the Cold War?" From a review of the relevant literature on covert action, international relations, and democracies, the following model was produced: Survival (s) = Leader Characteristics (lc) + Public Dissent (Pd) + Global Instability (gi) + Internal Conflict (IC) + Regime Type (RT) or s= lc+Pd+gi+IC+RT. In addition to the hypothesis that, collectively, these factors determine the outcome of leader survival, it was further proposed that leadership characteristics and regime type would have the most influence on an outcome.

As stated in the methodology section, fsQCA is applicable in this research because it remains relatively uncommon in international relations research. However, its potential holds much promise. The foundation of QCA is the truth table that allows the researcher to list and analyze many possible combinations of variables prior to arriving at the outcome. Ragin and Rihoux outline the steps to creating a truth table: "First, construct the truth table based on theoretical and case study information. Second, analyze the truth table. Finally, take the results back to the cases and theory to evaluate the findings" (2004, 4). Employing fsQCA this researcher assessed how various constellations of leader characteristics, public dissent, global instability, internal conflict, and regime type affected leader survival after covert action events.

Leader characteristics (LDR) sets the foundation for the case study in relation to the leader and the political and socio-cultural environments in the country at the time as well as any relationships the leader had with American or other mentors, confidantes, decision-makers, or advisors. Details captured in declassified national security documents and psychological profiles of leaders gleaned from CIA documents and personal biographies contributed to the assessment that those who had more agreeable personalities and enjoyed close ties with the United States may have experienced a more favorable fate. Conversely, leaders who did not enjoy close relations with US advisors, who had disagreeable, violent, and irrational temperaments, tended to amplify the political and socio-cultural situation in the country at the time, thereby resulting in a less favorable fate.

Public dissent (DISS) explores the existence of protests or mass mobilization of local populace against the leader at the time. The existence of mass protests or popular dissatisfaction with the leader indicates a weak support network for the individual. Intense tribal rivalries, hatred of minorities, discrimination and violence toward ethnic minorities, and any type of social injustice are fuel to the fire for mass social movements against a leader and their administration.

Global instability (INS) indicates that the country endured war or conflict that may have altered its future course and demonstrates the influence of powerful political actors: the United States represented democratic values and freedom while the Soviet Union's communist doctrine espoused socialism at all levels of society. A leader's position in either camp was an indicator of a favorable or unfavorable fate. In some cases, such as that of Indonesia, a leader advocated for democratic values but had intentions to ground it in communist doctrine.

Internal conflict (CON) indicates that the country has experienced or is experiencing a coup and/or revolution. Constant churning of political turmoil weakens the country's foundation and creates chaos at all levels. In addition, internal conflict wreaks havoc on a country's socio-economic conditions and eliminates it from being a competitor in the global economy.

In many of the case studies under consideration in this study, including Iran, Guatemala, Chile, and the Democratic Republic of the Congo, the country involved had abundant amounts of natural resources that were seized or exploited by external forces. Actions by leaders to regain these natural resources more often than not led to the leader's unfavorable fate.

Regime type (REG) describes the type of political system in place during the time of the covert action event. Iran was a parliamentary political system operating as a fledgling democracy; Guatemala was a praetorian-centric authoritarian regime that placed a tremendous amount of authority and influence on the military; Indonesia was a communist regime thinly veiled as a pseudo-democracy; Chile had a reputation for being a long-standing democracy until the election of Salvador Allende; the Democratic Republic of the Congo was a tribal-based governance that had a passionate desire to become a democratically liberated country free from colonialism's brutality; the Dominican Republic's decades-long reign of terror at the hands of a brutal dictator; the Republic of Vietnam was divided between the communists led by Ho Chi Min in the north and a puppet president reluctant to risk his safety to pursue democracy for his people in the south; and Cuba never pretended to be anything more than it was. Castro lived, breathed, and died a communist leader. Surprisingly, the type of regime had a minimal effect on a leader's favorable or unfavorable fate. Table 11.1 offers a visual summary of the relationships of the variables in relation to the factors of the model and their outcome. It also demonstrates the different combinations that can be used in the QCA methodology.

Table 11.1.

Country	*Leader Characteristics (LDR)*	*Public Dissent (DISS)*	*Global Instability (INS)*	*Internal Conflict (CON)*	*Regime Type (REG)*	*Outcome*
Iran	Mostly In	More In Than Out	Fully In	Fully In	Mostly Out	Successful
Guatemala	More Out Than In	More In Than Out	Fully In	Fully In	More Out Than In	Successful
Indonesia	Fully In	Mostly In	More In Than Out	Fully In	Mostly Out	Unsuccessful
Dominican Republic	Fully In	Fully In	More In than Out	More Out Than In	Mostly Out	Successful
Democratic Republic of the Congo	Fully In	Mostly Out	More Out Than In	Fully In	Fully In	Successful
Cuba	Fully In	Mostly In	Fully In	More Out Than In	More Out Than In	Unsuccessful
Vietnam	Mostly Out	Mostly In	Mostly In	Mostly In	More Out Than In	Successful
Chile	Fully In	Mostly In	Fully In	Mostly In	More In Than Out	Successful

SUCCESSFUL COVERT ACTION EVENTS AND LEADER SURVIVAL

Most successful covert action events demonstrate a general adherence to the factors of the model and indicate the validity of the model. The exceptions to this observation indicate a potential relationship between the factors of the model, combinations of factors, and a successful outcome. All of the successful cases, except for Guatemala and the Republic of Vietnam, demonstrated a Fully In or Mostly In degree of membership with the factor of leadership characteristics as predicted by the hypothesis. In the successful cases that did not display a strong membership with this factor, they did appear to display a strong adherence to global instability and internal conflict. In addition to indicating that successful cases do not necessarily adhere to all the principles of the model, these cases point to a relationship between leader characteristics, global instability, and internal conflict. Specifically, in cases where leadership characteristics are not as strong, external influences creating global instability, as well as internal conflict, resulted in a successful covert action event. The other general observation from this initial review of the data is the importance of leadership characteristics and their effects on regime type. This result indicates the reasonable assumption that a strong leader personality profile is significant in predicting leader survival.

It is unclear if this factor is unimportant or in cases with strong adherence to leadership characteristics, the leaders have strong ties to external governments or strong support from the local populace. In addition, all the cases demonstrated more in than out, mostly in, and fully in adherence. This result confirmed that public dissent is an influential variable in determining leader survival.

FAILED COVERT ACTION EVENTS AND LEADER SURVIVAL

The failed case studies were not too remarkably different from the successful case studies. Both failed case studies demonstrated a strong adherence to leader characteristics, public dissent, global instability, and internal conflict. Additionally, they showed a weak adherence to regime type. An interesting note with these case studies is that they demonstrated different relationships to external forces. In the case of Indonesia, Sukarno had a positive relationship with the United States and enjoyed support in his guided democracy vision. In the case of Cuba, Castro did not enjoy a positive relationship with the United States and was aggressive in his support to building Cuba's communist regime.

As with the successful case studies, the lack of adherence to the factors of the model does not imply causation. However, the failed case studies do suggest a correlation between leader characteristics, relationships with the United States, and leader survival.

It is important to note that this model is a preliminary proposal based on a minimum list of critical conditions. These conditions can be mixed in a variety of combinations. However, the main point to make in this research is that the conditions do not work in isolation to create the result. To create this idea, the researcher was careful to choose case studies that included the same silver-lined threading through them (US-sanctioned covert action) but was unfamiliar with all of the case studies until the research process began. This approach ensured that the research could capture all the details during the discovery phase to discover case-specific configurations that explain how leader survival is influenced by personality characteristics, public dissent, global instability, internal conflict, and regime type. The discovery phase was enjoyable, and Rihoux and Ragin describe the process clearly when they assert, "Surprising results then provide opportunities for further explorations to refine theory" (2009).

LEADER CHARACTERISTICS (LDR)

To measure charisma, neuroticism, extraversion, openness to experience, agreeableness, and conscientiousness, the researcher explored psychological profiles and transcripts from declassified sources as well as open-source biographies and articles. The resulting scale has a sample range from 0.6 (very high leader characteristics) to 0.01 (very low leader characteristics).

PUBLIC DISSENT (DISS)

Public dissent is measured by major episodes of political violence as documented by the State Fragility Matrix and Center for Systemic Peace, Coups d'Etat, 1946–2018. The 1953 ouster of Iran's Mossadegh government by the shah's loyalists resulted in 800 deaths. Civil violence experienced in Guatemala under the Arbenz regime resulted in one thousand deaths. The 1957–1961 Indonesia civil violence, attributed to dissident military members, resulted in thirty thousand deaths. There were no major episodes of political violence listed for the Democratic Republic of the Congo from 1960–1961. Chile's civil violence in 1973 as a result of the army's ousting of Allende resulted in five thousand deaths. No major episodes of political violence were listed for the Dominican Republic from 1959–1961, and no major episodes of political violence were listed for the Republic of Vietnam in 1963. Finally, the 1961–1962 Bay of Pigs invasion in Cuba resulted in 700 deaths. The resulting scale has a sample range from 0.6 (very high level of public dissent) to 0.01 (very low level of public dissent).

GLOBAL INSTABILITY (INS)

Global instability is captured by the major political violence events occurring during the time of the event. The Worldwide Governance Index (WGI) developed by Kaufmann et al. (2011) combine political independence, the quality of policy formulation and implementation, into an index for government effectiveness that ranges from -2.5 to +2.5.

INTERNAL CONFLICT (CON)

Iran

The Iranian Revolution of 1979 and the abolition of the monarchy occurred at the same time as the new Islamic Republic of Iran that combined elements of a parliamentary democracy with a Shiite Islamic theocracy. Since 1946, Iran had been involved in both interstate and intrastate conflicts.

Guatemala

Dominated by the conservative landholding class, aristocratic rule created an unstable political climate with military coups and conflicts along social class lines. Human rights abuses at the hands of government-affiliated death squads were the trademark for the regime.

Indonesia

Political infighting, undisputed territory, and an Indonesian government that was undecided between democracy and communism created a fertile environment for volatile Islamic movements, leading to intrastate conflict over governmental control. A narcissist leader with a leaning toward socialist doctrine prevented the country from achieving a stable form of governance.

Dominican Republic

The Dominican Republic became a sovereign state in 1865. Governance in the twentieth century vacillated between political upheavals and autocracy. The era of dictatorship was dipped in a prolonged bloodbath that transcended three decades under Rafael Trujillo's rule.

Democratic Republic of the Congo

Tribal infighting, inexperienced political leaders, and a strong collective memory of brutal colonial rule culminated in a string of presidents all elected under unconstitutional means.

Cuba

Castro's regime became increasingly anti-American as its economic reforms collided with American interests in the country. The reforms also led to the exodus of thousands of upper- and middle-class Cubans, primarily to the United States.

Republic of Vietnam

The Democratic Republic of Vietnam in the north controlled by the communist Vietminh led by Ho Chi Minh and the Republic of Vietnam in the south led by a US-supported Diem.

Chile

Allende's narcissist leadership style ultimately resulted in a coup against his democratically elected Socialist party.

REGIME TYPE (REG)

Iran's focus is on independence; Guatemala's focus is on military governments; Indonesia's focus is on its nationalist party; Democratic Republic of the Congo's focus is on the Democratic Union for the Defense of African Interests; Chile's regime focus is on Salvador Allende and his military government; the Dominican Republic's focus is on Rafael L. Trujillo; the Republic of Vietnam's focus is on military governments; and Cuba's focus is on Fidel Castro's government.

Through a review of the relevant doctrine in the literature, the researcher predicted that leader characteristics would be the most influential factor in the model. The argument for the importance of leadership characteristics is logical. Charismatic leadership can overcome the most difficult situations, and this is the reason that leader characteristics was determined to be the most important factor.

The connection between leader characteristic and regime type is logical. Leaders are tied to the institution they lead, which in turn, drives internal and external responses. Due to the secrecy and violent nature of covert action events and the necessity to keep the information from public scrutiny, Global Instability turned out to be one of the most influential factors in the case studies. American leaders assumed that communism was the greatest threat, and the United States would eliminate it at all costs. The long-term consequences of these covert action events are seen today. Public dissent and internal conflict were also important factors in leader survival.

Subset/Superset Analysis

To determine the coverage and consistency, the next step was to analyze the cases using the superset/subset function of the fsQCA program. This procedure analyzes how well the outcome, or dependent variable "success" rate, fits with the necessary conditions—in this case, the factors of the model that produced that outcome, as specified in Table 11.2.

Table 11.2.

Factor	*Consistency*	*Coverage*	*Combined*
Leader Characteristics (LDR)	0.961493	0.817518	0.703680
Public Dissent (DISS)	0.745314	0.717693	0.887707
Global Instability (INS)	0.920097	0.762431	0.920139
Internal Conflict (CON)	0.663362	0.615693	0.698630
Regime Type (REG)	0.617161	0.612377	0.653244

ANALYSIS OF NECESSARY CONDITIONS

An analysis of necessary conditions allows researchers with another tool to measure consistency and coverage. In this case, "consistency assesses the degree to which instances of the outcome agree in displaying the causal condition thought to be necessary, while coverage assesses the relevance of the necessary condition—the degree to which instances of the condition are paired with instances of the outcome" (Ragin 2008, 44–45). This analysis provides insight into potential interactions among factors in the model and the influence of adherence to one factor relative to adherence to others.

The analysis of necessary conditions among successful cases provided some validation of the model and some differences. If the model is correct, leadership characteristics, public dissent, and global instability should have the highest consistency scores. Interestingly, regime type demonstrated the lowest coverage score, indicating that it is a necessary condition but not sufficient to produce a successful outcome on its own.

The necessary condition analysis also produced other interesting coverage results. The ranking among the factors from most to least significant was leader characteristics, public dissent, internal conflict, global instability, and regime type. The high score for public dissent is possibly because the factor is inherent to the political-social element and social movements. It is also possible that this factor is dependent on the duration of the leader's tenure as well as divisions within the population based on ethnicity, economic status, and literacy rate among others. All of the successful cases considered unfolded during the Cold War context, so by necessity, all of the programs had to be designed to be as self-sufficient as possible.

Table 11.3.

Factor	*Consistency*	*Coverage*
Leader Characteristics (LDR)	0.812030	0.962138
Public Dissent (DISS)	0.630150	0.626753
Global Instability (INS)	0.714286	0.957179
Internal Conflict (CON)	0.810150	0.962054
Regime Type (REG)	0.440226	0.467961

Leader characteristics and public dissent proved to be the most critical factors in producing failure. The other finding of note was the lack of adherence to regime type. This result again suggests that this factor is a necessary, but not sufficient condition.

Table 11.4.

Factor	*Consistency*	*Coverage*
Leader Characteristics (LDR)	.173913	1
Public Dissent (DISS)	.086957	1
Global Instability (INS)	.086957	1
Internal Conflict (CON)	.086957	1
Regime Type (REG)	0	0

FUZZY-SET CALIBRATION

FsQCA allows for this phenomenon through a process of "calibration" (Ragin 2008, 16–17; 85–96). Informed by the available literature and familiarity with the sample of cases, calibration allows for individual factors to be weighted to account for their overall importance and influence on the other factors and to mitigate overrepresented factors. In the calibration process, to structure the fuzzy-set, each factor is assigned three qualitative breakpoints: full membership, cross-over point, and threshold for non-membership.

Fuzzy-Set Analysis

Using the calibrated values, a fuzzy-set analysis of necessary conditions was conducted. The results shown in Table 11.5 supported the model's prediction that leader characteristics and global instability are the two of the most necessary and sufficient factors in determining a successful outcome. Interestingly, incorporation of public dissent ranked the third highest in coverage. This could be because adherence to this factor in the cases was usually an existing undercurrent of social unrest based on socio-cultural factors. The results also indicate that leader characteristics and internal conflict are influenced by the other factors and socio-cultural conditions common to all the cases.

Table 11.5.

Factor	*Consistency*	*Coverage*
Leader Characteristics (LDR)	0.761278	0.959716
Public Dissent (DISS)	0.868421	0.700585
Global Instability (INS)	0.855263	0.899210
Internal Conflict (CON)	1.000000	0.886667
Regime Type (REG)	0.435667	0.457578

Analysis and Comparison of Results

The hierarchy predicted by the model and the crisp and fuzzy-set analyses were consistent in the finding that leader characteristic is the most critical factor in determining a successful outcome. This finding is intuitive, but also informative for policymakers in determining whether to engage in covert action strategy. While less costly than executing large forces, unconventional forces that execute covert action are inexpensive in labor force or material, comparatively speaking. The other critical factors identified by the model, internal conflict and regime type did not fare as well in crisp set analysis. Interestingly, after calibration the scores for internal conflict and regime type both experienced lower scores as it became apparent that they were highly influenced by other factors in the model.

Cross-Case Comparative Analysis

According to Stretton (1969, 158), "Cross-case analysis examines the commonalities and difference in the events, activities, and processes of case studies. Additionally, it is conducted in order to 'extend the investigator's experience beyond the single case by provoking her or his imagination, prompting new questions, seeking new dimensions, measuring alternatives, creating models, and constructing ideal types and utopias.'" Ragin (1997) adds, "Another purpose of engaging in cross-case analysis is to enhance researchers' capacity to understand 'how' relationships may exist among discrete cases, accumulate knowledge on the original case, refine or develop concepts" (Ragin 1997).

Fuzzy-Set Truth Table Analysis

The final analysis conducted was the fuzzy-set truth table analysis. The function of this analysis is to "determine the connections between combinations of causal conditions and an outcome" (Ragin 2008, 109). The standard analysis provides three sets of solutions: complex, parsimonious, and intermediate. The complex case provides a solution that avoids counterfactual cases. The parsimonious case provides the simplest case with the fewest casual conditions through the use of counterfactual cases. The intermediate case allows for inclusion of the most reasonable counterfactual cases. The strength of this analytical method is it determines multiple pathways to a similar outcome.

The parsimonious solution includes two causal pathways: leader characteristics and internal conflict. The overall solution was calculated to have coverage of 0.998175 and a consistency of 0.868254. In this solution, each term is measured by their degree membership in each possible solution for the outcome, and the proportion of each factor contributes to the explanatory value in each potential solution. The importance of this finding to policymakers is that understanding a leader's personality profile is an important consideration in designing a covert action strategy.

The intermediate solution provided causal pathways leading to a successful outcome. The first is leader survival, public dissent, and global instability. The second is internal conflict and regime type. These pathways provide an overall consistency of 0.875614 and an overall coverage of 0.976277. These two pathways are instructive in that they indicate two realistic sets of conditions that can produce a successful outcome. In cases where leader characteristics are strong, regime type was not clearly defined, and where internal conflict is strong, the more advantageous the opportunity to execute a successful covert action event. Public dissent is also another factor that influences the probability for success in creating a permissive environment. This research contributes to the study of covert action literature in three ways. First, it analyzes political and socio-cultural conditions that existed in the countries. Previous works have focused on these conditions from a US perspective. Second, contrary to most Cold War–era covert action studies, this book focuses on leader personality profiles within the context of the environment at that time. Finally, this research focuses on factors influencing leader survival, whereas previous research focused on the CIA's use of covert action, foreign policy from US presidents' perspective, and, more often than not, the failures of these events and their exploitation as depicted by popular culture and media at the time. Covert action will continue to exist, and US presidents will need to know how to execute it smartly. This research also shows the limits of covert action, the nuances of the intelligence tradecraft, the impact of US foreign policy on developing countries, and the long-lasting effects of those decisions. Additional research may include a more in-depth look at the correlation between leader characteristics and other covert action events in history.

Bibliography

Abrahamian, Ervand. 2001. "The 1953 Coup in Iran." *Science & Society* 65(2) (Summer): 182–215.

Acemoglu, D., S. Johnson, and J. Robinson. 2006. "Institutions as the Fundamental Cause of Long Run Growth." In *Handbook of Economic Growth*, ed. R. Aghion and S. Durlauf. Oxford: Oxford University Press, 385–472.

Acemoglu, Daron, Simon Johnson, and James A. Robinson. 2001. "The Colonial Origins of Comparative Development: An Empirical Investigation." *American Economic Review,* 91(5): 1369–1401.

Ahern, Thomas L. 1998. *CIA and the Generals: Covert Support to Military Government in South Vietnam – Secret.* Washington, DC: Central Intelligence Agency, Center for the Study of Intelligence.

Aksoy, Deniz; David B. Carter, and Joseph Wright. 2015. "Terrorism and the Fate of Dictators." *World Politics* 67(3): 423–468.

Alexander L. George and Andrew Bennett. 1993. *Bridging the Gap: Theory & Practice in Foreign Policy.* Washington, DC: United States Institute for Peace Press.

Alexander L. George and Andrew Bennett. 2005. *Case Studies and Theory Development in the Social Sciences.* Cambridge, MA: MIT Press.

Allison, Graham. 1969. "Conceptual Models and the Cuban Missile Crisis." *American Political Science Review* 63 (September): 689–718.

Allison, Paul. 1984. *Event History Analysis.* Beverly Hills, California: Sage.

Allison, Roy. 1988. *The Soviet Union and the Strategy of Non-alignment in the Third World.* Cambridge, UK: Cambridge University Press.

Allyn, Bruce J., James G. Blight, and David A. Welch. 1989/90. "Essence of Revision: Moscow, Havana, and the Cuban Missile Crisis." *International Security* 14(3) (Winter): 144–46.

Amenta, Edwin, and Jane D. Poulsen. 1994. "Where to Begin: A Survey of Five Approaches to Selecting Independent Variables for Qualitative Comparative Analysis." *Sociological Methods & Research* 23(1): 22–53.

Anderson, David L. 2004. *Columbia Guide to the Vietnam War*. New York: Columbia University Press.

Anderson, Christopher J. and Christine A. Guillory. 1997. "Political Institutions and Satisfaction with Democracy: A Cross-National Analysis of Consensus and Majoritarian Systems. *The American Political Science Review* 91(1) (March): 66–81.

Armingeon, K. et al. 2012. *Comparative Political Data Set III, 1990–2010*. Bern, Switzerland: Institute of Political Science, University of Bern.

Armony, Ariel and H. Schamis. 2005. "Babel in Democratization Studies." *Journal of Democracy* 16(4) (October): 113–128.

Arquilla, John. 2006. *The Reagan Imprint: Ideas in American Foreign Policy from the Collapse of Communism to the War on Terror*. Chicago: Ivan R. Dee.

Art, Robert. 1996. "American Foreign Policy and the Fungibility of Force." *Security Studies* 5 (Summer): 7–42.

Axelrod, Robert and Robert O. Keohane. 1985. "Achieving Cooperation under Anarchy: Strategies and Institutions." *World Politics* 38: 226–254.

Axelrod, Robert. 1984. *The Evolution of Cooperation*. Ch. 1–4. New York: Basic Books.

Baaz, Maria Eriksson, and Maria Stern. 2011. "Whores, Men, and Other Misfits: Undoing Feminization in the Armed Forces in the DRC," *African Affairs* 110(441): 563–5.

Baldwin, David A., ed. 1993. *Neorealism and Neoliberalism: The Contemporary Debate*. New York: Columbia University Press.

Barkin, Samuel J. and Bruce Conin. 1994. "The State and the Nation: Changing Norms and the Rules of Sovereignty in International Relations." *International Organization* 48(1): 107–130.

Barnett, Michael and Raymond Duvall. 2005. "Power in International Politics." *International Organization* 59 (Winter): 39–75.

Barron, Thomas. 2013. "The Soldier and the State in the Congo Crisis: The Unprofessional Legacy of the National Congolese Army" *African Security* 6(97): 98–132.

Bass, B. M. 1985. *Leadership and Performance Beyond Expectations*. New York: Free Press.

_____. 1990. *Bass and Stogdill's Handbook of Leadership* (3 ed.). New York: Free Press.

_____. 1995. "Theory of Transformational Leadership Redux." *Leadership Quarterly* 6: 463–478.

Bates, Robert. 2008. *When Things Fall Apart*. Cambridge, UK: Cambridge University Press.

Baumgartner, Michael. 2015. "Parsimony and Causality." *Quality & Quantity* 49(2), 839–856.

Beck, Nathaniel. 2010. "Time Is Not a Theoretical Variable." *Political Analysis* 18(3): 293–294.

Beck, Colin J. 2014. "Reflections on the Revolutionary Wave in 2011." *Theory and Society* 43(2) (March): 197–223.

Belkin, Aaron and Evan Schofer. 2003. "Toward a Structural Understanding of Coup Risk." *Journal of Conflict Resolution* 47(5): 594–620.

_____. 2005. "Coup Risk, Counterbalancing, and International Conflict." *Security Studies* 14(1): 140–177.

Bemis, Samuel Flagg. 1943. *The Latin American Policy of the United States*. New York: Harcourt, Brace & Co.

Benedict, John and Keith T. Poole. 1996. "Does High Income Promote Democracy?" *World Politics* 49(1): 1–30.

Bennett, Andrew and Colin Elman, 2007. "Case Study Methods in the International Relations Subfield." *Comparative Political Studies*. 40(2): 170–195.

Bennett, S. 1996. "Security, Bargaining and the End of Interstate Rivalry." *International Studies Quarterly* 40: 157–84.

Bennett, D. Scott, Allan C. Stam. 1998. "The Declining Advantages of Democracy." *The Journal of Conflict Resolution* (June): 42–43.

Berger, Andreas, Cassy L. Dorff, and Michael D. Ward. 2014. "Ensemble Forecasting of Irregular Leadership Change." *Research and Politics* (October–December): 1–7.

Berger, Thomas. 1998. *Cultures of Antimilitarism: National Security in Germany and Japan*. Baltimore, MD: Johns Hopkins Press.

Berg-Schlosser D., G. de Meur, B Rihoux, and C. C. Ragin. 2009. "Qualitative Comparative Analysis as an Approach." In *Configurational Comparative Methods. Qualitative Comparative Analysis (QCA) and Related Techniques*, edited by B. Rihoux and C. C. Ragin, 1–18. Los Angeles: Sage.

Berg-Schlosser, Dirk, Gis`ele de Meur, Charles C Ragin, and Bentoit Rihoux. 2008. *Qualitative Comparative Analysis as an Approach*. Thousand Oaks: Sage.

Berkowitz, Bruce. 1998. "The Logic of Covert Action." *The National Interest* (Spring): 1–10.

Beschloss, Michael. 1991. *The Crisis Years: Kennedy and Khrushchev, 1960–1963*. New York: Edward Burlingame.

Betts, Richard. 1982. *Surprise Attack: Lessons for Defense Planning*. Washington DC: Brookings Institute.

Biddle B. J. 1979. *Role Theory: Expectations, Identities, and Behavior*. New York: Academic Press.

Biddle, Stephen and Robert Zirkle. 1996. "Technology, Civil–military Relations, and Warfare in the Developing World." *Journal of Strategic Studies* 19(2): 171–212.

Bienen, Henry and Nicolas van de Walle. 1989. "Time and Power in Africa." *The American Political Science Review* 83: 19–34.

_____. 1991. *Of Time and Power: Leadership Duration in the Modern World*. Stanford, CA: Stanford University Press.

_____. 1992. "A Proportional Hazard Model of Leadership Duration." *The Journal of Politics* 54(3) (August): 685–717.

Bissell, Richard M. 1996. *Reflections of a Cold Warrior: From Yalta to the Bay of Pigs*. New Haven, CT: Yale University Press.

Black, George. 1988. *The Good Neighbor: How the United States Wrote the History of Central America and the Caribbean*. New York: Pantheon Books.

Black, Jan Knippers. 1986. *The Dominican Republic: Politics and Development in an Unsovereign State*. London: Unwin Hyman.

Blair, Robert A., Christopher Blattman, and Alexandra Hartman. 2017. "Predicting Local Violence: Evidence from a Panel Survey in Liberia." *Journal of Peace Research* 54(2): 298–312.

Block, J. 1995. "A Contrarian View of the Five-Factor Approach to Personality Description." *Journal of Personality and social Psychology* 117: 187–215.

Blum, William. 2006. *Rogue State: A Guide to the World's Only Superpower.* Monroe, ME: Common Courage Press.

Booth, Ken. 1991. "Security in Anarchy: Utopian Realism in Theory and Practice." *International Affairs* 67(3): 527–545.

Boix, Carles and Milan Svolik. 2013. "The Foundations of Limited Authoritarian Government: Institutions, Commitment, and Power-Sharing in Dictatorships." *Journal of Politics* 75(2): 300–316.

Bowen, Gordon L. 1983. "U.S. Foreign Policy toward Radical Change: Covert Operations in Guatemala, 1950–1954." *Latin American Perspectives* 10(1): 88–102.

Brands, Hal. 2011. "Crime, Irregular Warfare, and Institutional Failure in Latin America: Guatemala as a Case Study." *Studies in Conflict & Terrorism* 34(3): 228–247.

Brands, H. W. 1983. "The Limits of Manipulation: How the United States Didn't Topple Sukarno." *The Journal of American History* 76(3) (December): 789.

Brantly, Aaron. 2012. "On the Utility of Covert Action." APSA 2012 Annual Meeting Paper. Available at SSRN: https://ssrn.com/abstract=2110871.

Bratton, Michael. 2009. "Sub-Saharan Africa." In *Democratization,* edited by Christian W. Haerpfer, Patrick Bernhagen, Ronald F. Inglehart, and Christian Welzel. New York: Oxford University Press.

Bratton, Michael and Nicolas van de Walle. 1994. "Patrimonial Regimes and Political Transitions in Africa." *World Politics* 46(4): 453–489.

Breen, John G. 2016. "Disciples: The World War II Missions of the CIA Directors Who Fought for Wild Bill Donovan." *Military Review* 96(3) (May/June): 126–128.

Briggs, Ellis O. 1964. *Farewell to Foggy Bottom: The Recollections of a Career Diplomat.* New York: David McKay.

Brockett, Charles D. 2002. "An Illusion of Impotence: U.S. Policy toward Guatemala, 1954–1960." *Latin American Politics and Society* 44(1) (Spring): 91.

Brooks, Stephen G. 1997. "Dueling Realisms." *International Organization* 51(3) (Summer): 445–77.

Brown, Michael E. 1996. *Debating the Democratic Peace*. Boston, MA: The MIT Press.

_____. 2004. *Offense, Defense, and War*. Boston, MA: The MIT Press.

Buell, Hal. 2013. "Capturing the Ultimate Protest." *Vietnam* 26(1) (June): 40–45.

Bueno de Mesquita, Bruce and David Lalman. 1992. *War and Reason: Domestic and International Imperatives.* New Haven, CT: Yale University Press.

Bueno de Mesquita, Bruce and Kenneth Shepsle. 2003. "Predicting Politics." *Political Science Quarterly* 118(2) (June): 324–326.

Bueno de Mesquita, Bruce and Randolph M. Siverson. 1995. "War and the Survival of Political Leaders: A Comparative Study of Regime Types and Political Accountability." *American Political Science Review* 89: 841–55.

Bueno de Mesquita, Bruce, Randolph M. Siverson, and Gary Woller. 1992. "War and the Fate of Regimes: A Comparative Analysis." *The American Political Science Review* 86(3) (September): 638–646.

Bueno de Mesquita, Bruce and Alastair Smith. 2004. "Testing Novel Implications from the Selectorate Theory of War." *World Politics* 56 (April): 363–88.

_____. 2009. "Political Survival and Endogenous Institutional Change." *Comparative Political Studies* 42(2): 167–97.

_____. 2010. "Leader Survival, Revolutions, and the Nature of Government Finance." *American Journal of Political Science* 54(4): 936–50.

Bueno de Mesquita B. et al. 2005. *The Logic of Political Survival*. Cambridge, MA: MIT Press.

Bull, Hedley. 1977. *The Anarchical Society: A Study of Order in World Politics*. London: Macmillan.

Bundy, McGeorge. 1961a. Memorandum of Discussion on Cuba. Foreign Relations of the United States, 1961–1963, Volume X, Cuba, January 1961–September 1962, Document 30. Washington, D.C.: Office of the Historian, U.S. Department of State, 28 Jan. 1961. https://history.state.gov/historicaldocuments/frus196163v10/d30. 22 Nov. 2016.

_____. 1961b. Memorandum from the President's Special Assistant for National Security Affairs (Bundy) to President Kennedy. Foreign Relations of the United States, 1961–1963, Volume X, Cuba, January 1961–September 1962, Document 64. Washington, D.C.: Office of the Historian, U.S. Department of State, 15 Mar. 1961. https://history.state.gov/historicaldocuments/frus196163v10/d64. 7 Jan. 2017.

Burns, James MacGregor. 1956. *Roosevelt: The Lion and the Fox*. New York: Harcourt, Brace & World.

Buzan, Barry. 1984. "Peace, Power, and Security: Contending Concepts in the Study of International Relations." *Journal of Peace Research* 21(2): 109–125.

_____. 1993. "From International System to International Society: Structural Realism and Regime Theory Meet the English School." *International Organization* 47(3) (Summer): 327–352.

_____. 2001. "The English School: An Underexploited Resource in IR." *Review of International Studies* 27(3): 471–88.

Byman, Daniel and Matthew Waxman. 2002. *The Dynamics of Coercion: American Foreign Policy and the Limits of Military Might*. Cambridge, UK: Cambridge University Press.

Byman, D. 2005. *Deadly Connections: States That Sponsor Terrorism*. Cambridge, UK: Cambridge University Press.

Cabot, John Moors. 1979. *First Line of Defense: Forty Years' Experience of a Career Diplomat*. Washington, DC: School of Foreign Service, Georgetown University.

Calder, Bruce. 1984. *The Impact of Intervention: The Dominican Republic during the U.S. Occupation of 1916–24*. Austin, TX: Austin University Press.

Calhoun, Frederick. 1986. *Power and Principle: Armed Intervention in Wilsonian Foreign Policy*. Kent, OH: Kent State University.

Callaghy, Thomas M. 2001. "Life and Death in the Congo: Understanding a Nation's Collapse." *Foreign Affairs* 80(5) (September/October).

Callahan, James. 2010. *Covert Action in the Cold War: US Policy, Intelligence and CIA Operations*. London: I. B. Tauris.

Carothers, Thomas. 1991. *In the Name of Democracy: U.S. Policy toward Latin America in the Reagan Years.* Berkeley, CA: University of California Press.

Carr, Edward Hallett. 1939. *The Twenty Years' Crisis, 1919–1939: An Introduction to the Study of International Relations*. New York: St. Martin's Press.

Carter, John J. 2000. *Covert Operations as a Tool of Presidential Policy in American History from 1800–1920.* Lewiston, NY: The Edwin Mellen Press.

_____. 2006. *Covert Action as a Tool of Presidential Foreign Policy: From the Bay of Pigs to Iran-Contra.* Lewiston, NY: The Edwin Mellen Press.

Cassidy, Robert M. 2004. "Back to the Street without Joy: Counterinsurgency Lessons from Vietnam and Other Small Wars." *Parameters* (Summer): 74–83.

Castro, Fidel and Ignacio Ramonet. 2007. *Fidel Castro: My Life: A Spoken Autobiography*. Translated by Allen Lane. Great Britain: Penguin Books.

Center for International Development and Conflict Management (CID CM). 2006. Minorities at Risk Project, http: //www.cidcm.umd.edu/inscr/mar/.

Center for the Study of Civil War, PRIO. 2006. Armed Conflict Data Version 4-2006b. http://www.prio.no/CSCW/Datasets/Armed-Conflict/UCDP-PRIO/Old-Versions/4-2006b/.

Central Intelligence Agency. 1961a. Paper Prepared in the Central Intelligence Agency. In: Foreign Relations of the United States, 1961–1963, Volume X, Cuba, January 1961–September 1962, Document 58. Washington, DC: Office of the Historian, U.S. Department of State, 11 Mar. 1961. https://history.state.gov/historicaldocuments/frus1961-63v10/d58 (accessed: 7 Jan. 2017).

_____. 1961b. Revised Cuban Operation. In: Foreign Relations of the United States, 1961–1963, Volume X, Cuba, January 1961–September 1962, Document 61. Washington, D.C.: Office of the Historian, U.S. Department of State, 15 Mar. 1961. https://history.state.gov/historicaldocuments/frus196163v10/d61 (accessed: 7 Jan. 2017).

_____. 2016. "Central Intelligence Agency's Freedom of Information Act Electronic Reading Room." Last Modified May 16, 2016. https://www.cia.gov/library/readingroom/.

_____. NIC Files, Job 79S01011A, Box 3, Folder 6, SE–6 Current Developments in Iran.

Champion, Brian. 2008. "Spies (Look) Like Us: The Early Use of Business and Civilian Covers in Covert Operations." *International Journal of Intelligence and Counterintelligence* 21(3) (2008): 530–564.

Chan K. Y. and Drasgow F. 2001. "Toward a Theory of Individual Differences and Leadership: Understanding the Motivation to Lead." *Journal of Applied Psychology* 86: 481–498.

Cheibub, Jose Antonio, Jennifer Gandhi, and James Raymond Vreeland. 2010. "Democracy and Dictatorship Revisited." *Public Choice* 143(1): 67–101.

Cherulnik P. D., Turns L. C., Wilderman S. K. 1990. "Physical Appearance and Leadership: Exploring the Role of Appearance-Based Attribution in Leader Emergence." *Journal of Applied Social Psychology* 20: 1530–1539.

Chester L. Cooper, recorded interview by Joseph E. O'Connor, May 6, 1966, John F. Kennedy Library Oral History Program.

Chiozza, Giacomo, and Henk E. Goemans. 2003. "Peace through Insecurity: Tenure and International Conflict." *Journal of Conflict Resolution* 41(4): 443–467.

_____. 2004. "International Conflict and the Tenure of Leaders: Is War Still Ex Post Inefficient?" *American Journal of Political Science* 48(3): 604–619.

Church Committee (US Senate Select Committee to Study Governmental Operations with Respect To Intelligence Activities). 1975. *Alleged Assassination Plots Involving Foreign Leaders, 1975 US Senate Report on CIA Covert Operations to Kill Fidel Castro, Ngo Dinh Diem, and Others*. Washington, DC: Government Printing Office.

Cohen, Herman J., Charles G. Cogan, and Stephen R. Weissman. 2015. "Who Lost Congo? The Consequences of Covert Action." *Foreign Affairs* 94(1): 169–173.

Coll, Steve. 2001. *Ghost Wars: The Secret History of the CIA, Afghanistan, and Bin Laden, From the Soviet Invasion to September 10, 2001.* New York: Penguin Books.

Collier, Paul. 2003. *Breaking the Conflict Trap: Civil War and Development Policy.* Oxford, UK: Oxford University Press and Washington, DC: The World Bank.

Comments by Fidel Castro 14 and 15 June on the Invasion of 17 April 1961 (1961). Foreign Relations of the United States, 1961–1963, Volume X, Cuba, January 1961–September 1962, Document 236. Washington, DC: Office of the Historian, U.S. Department of State, 22 June 1961. https://history.state.gov/historicaldocuments/frus1961-63v10/d236. 22 Nov. 2016.

Conboy, Kenneth, and James Morrison. 1999. *Feet to the Fire: CIA Covert Operations in Indonesia, 1957–1958.* Annapolis, MD: Naval Institute Press.

Conclusions, CIA Clandestine Services History, Record of Paramilitary Action against the Castro Government of Cuba, 5th May 1961. http://www.foia.cia.gov/docs/DOC_0001459144/DOC_0001459144.pdf

Conger J. A. and R. N. Kanungo. 1987. "Toward a Behavioral Theory of Charismatic Leadership in Organizational Settings." *Academy of Management Review* 12: 637–647.

_____. 1992. "Perceived Behavioural Attributes of Charismatic Leadership." *Canadian Journal of Behavioural Science* 24: 86–102.

_____. 1994. "Charismatic Leadership in Organizations—Perceived Behavioral-Attributes and their Measurement." *Journal of Organizational Behavior* 15: 439–452.

Copeland, Dale C. 2003. "A Realist Critique of the English School." *Review of International Studies* 29(3) (July): 427–441.

Coppedge, Michael, Angel Alvarez, and Claudia Maldonado. 2008. "Two Persistent Dimensions of Democracy: Contestation and Inclusiveness." *Journal of Politics* 70 (July): 623–47.

Crassweller, Robert D. 1966. *Trujillo: The Life and Times of a Caribbean Dictator*. New York: McMillan Company.

Cribb, Robert. 2001. "Genocide in Indonesia, 1965–1966." *Journal of Genocide Research* 3(2): 219–239.

Cribb, Robert, and Charles A. Coppel. 2009. "A Genocide that Never Was: Explaining the Myth of anti-Chinese Massacres in Indonesia, 1965–66." *Journal of Genocide Research* 11(4): 447–465.

Croco, Sarah. E. 2011. "The Decider's Dilemma: Leader Culpability, War Outcomes, and Domestic Punishment." *American Political Science Review* 105(3): 457–77.

Cullather, Nick. 1999. *Secret History: The CIA's Classified Account of Its Operations in Guatemala 1952–1954.* Stanford, CA: Stanford University Press.

Cullather, Nicholas. 1994. *Operation PBSUCCESS: The United States and Guatemala, 1952–1954.* History Staff, Center for the Study of Intelligence, Central Intelligence Agency, Washington, DC.

Cumming, Alfred. 2006. *Covert Action: Legislative Background and Possible Policy Questions*. CRS Report for Congress.

Dabashi, Hamid. *Theology of Discontent: The Ideological Foundations of the Islamic Revolution in Iran*. New York: New York University Press.

Da Graca, John V. 2000. *Heads of State and Government*, 2nd Edition. New York: St. Martin's Press.

Dahl, Robert. 1971. *Polyarchy: Participation and Opposition*. New Haven, CT: Yale University Press.

_____. 1989. *Democracy and its Critics*. New Haven, CT: Yale University Press.

_____. 1998. *On Democracy.* New Haven: Yale University Press.

Daugherty, William J. 2004a. *Executive Secrets: Covert Action and the Presidency.* Lexington, KY: The University Press of Kentucky.

_____. 2004b. "Approval and Review of Covert Action Programs Since Reagan." *International Journal of Intelligence and Counterintelligence* 17: 62–80.

Davenport, Christian. 2007. *State Repression and the Domestic Democratic Peace.* Cambridge, UK: Cambridge University Press.

De Bellaigue, Christopher. 2012. *Patriot of Persia: Mohammad Mossadegh and a Tragic Anglo-American Coup*. New York: Harper Collins.

Department of State Publication 7326, *The Elements in Our Congo Policy*, December 1961.

Department of State (n.d.). Editorial Note. In: Foreign Relations of the United States, 1961–1963, Volume X, Cuba, January 1961–September 1962, Document 84.

DeRue D. S., S. J. Ashford, and N. Cotton. 2009. "Assuming the Mantle: Unpacking the Process by which Individuals Internalize a Leader Identity." In *Exploring Positive Identities and Organizations: Building a Theoretical and Research Foundation,* edited by L. M. Roberts and J. E. Dutton, 213–232. New York: Taylor & Francis.

Desch, Michael C. 2002. "Democracy and Victory: Why Regime Type Hardly Matters." *International Security* 27(2) (Fall): 5–47.

Deutsch, Karl W. 1961. "Social Mobilization and Political Development." *American Political Science Review* 55 (September): 493–514.

Devine, Jack. 2014. "What Really Happened in Chile: The CIA, the Coup Against Allende, and the Rise of Pinochet." *Foreign Affairs* 93(4): (July/August): 26–30, 31–35, 93.

De Witte, Ludo. 2001. *The Assassination of Lumumba*. London and New York:Verso Press.

de Soto, Jose M. Aybar. 1978. *Dependency and Intervention: The Case of Guatemala in 1954*. Boulder, CO: Westview.

Diamond, Larry. 2002. "Thinking about Hybrid Regimes." *Journal of Democracy* 13(2) (April): 21–35.

Doran, Charles F. and Wes Parsons. 1980. "War and the Cycle of Relative Power." *The American Political Science Review* 74(4) (December): 947–965.

Doyle, Mark. 2005. "Three Pillars of the Liberal Peace," *American Political Science Review* 99(3) (August): 463–466.

Doyle, Michael W. 1983. "Kant, Liberal Legacies, and Foreign Affairs." *Philosophy & Public Affairs* 12(3) (Summer): 205–235.

_____. 1986. "Liberalism and World Politics" *American Political Science Review* 80(4) (December): 1151–1169.

_____. 1997. *Ways of War and Peace: Realism, Liberalism, and Socialism*. New York: W. W. Norton & Company.

Downes, Alexander B. and Mary Lauren Lilley. 2010. "Overt Peace, Covert War?: Covert Intervention and the Democratic Peace." *Security Studies* 19: 2, 266–306.

Dulles, Allen. 1963. *The Craft of Intelligence*. London: Weidenfield and Nicholson.

Dunne, Michael. 2000. "US Foreign Relations in the Twentieth Century: From World Power to Global Hegemony." *International Affairs* 76(1): 25–40.

Dunning, Thad. 2005. "Resource Dependence, Economic Performance, and Political Stability." *Journal of Conflict Resolution* 49 (August): 451–82.

Easter, David. 2005. "'Keep the Indonesian Pot Boiling': Western Covert Intervention in Indonesia, October 1965–March 1966." *Cold War History* 1(5): 55–73.

Eden, D. 1992. "Leadership and Expectations: Pygmalion Effects and Other Self-Fulfilling Prophecies in Organizations." *Leadership Quarterly* 3: 271–305.

Efimenco, N. Marbury. 1955. "An Experiment with Civilian Dictatorship in Iran: The Case of Mohammed Mossadegh." *The Journal of Politics* 17(3) (August): 390–406.

Elandt-Johnson, R. C., and N. L. Johnson. 1980. *Survival Models and Data Analysis*. New York: Wiley.

Elman, Colin. 1996. "Horses for Courses: Why Not a Neorealist Theory of Foreign Policy?" *Security Studies* 6: 7–53.

Escalante, Fabian. 1995. *The Secret War: CIA Covert Operations in Cuba, 1959–62*. New York: Ocean Press.

Escribà-Folch, Abel. 2013. "Accountable for What? Regime Type, Performance, and the Fate of Outgoing Dictators, 1946–2004." *Democratization* 20(1): 160–85.

Escribà-Folch, Abel. 2014. "Repression, Political Threats, and Survival under Autocracy." *International Political Science Review* 34(5): 543–560.

Etcheson, Craig. 2005. *After the Killing Fields: Lessons from the Cambodian Genocide*. New York: Praeger.

Etges, Andreas. 2011. "All That Glitters Is Not Gold: The 1953 Coup against Mohammed Mossadegh in Iran." *Intelligence and National Security* 264(4): 95–508.

Fall, Bernard B. 1956. *The Vietminh Regime*. New York: Institute of Pacific Affairs.

_____. 1961. *Street without Joy*. Harrisburg, PA: The Stackpole Company.

_____. 967. *The Two Vietnams: A Political and Military Analysis* (2nd ed.). New York: Praeger.

Fearon, James D. 1995. "Rationalist Explanations for War." *International Organization* 49(3): 379–414.

Fearon, James D. and David D. Laitin. 2003. "Ethnicity, Insurgency, and Civil War." *American Political Science Review* 97(1): 75–90.

Feierabend, Ivo K., Rosalind L. Feierabend, Betty A. Nesvold, and Franz M. Jaggar. 1971. "Political Violence and Assassination: A Cross-National Assessment." In *Assassinations and the Political Order*, ed. William J. Crotty, 54–140. New York: Harper and Row.

Felix, Christopher. 2001. *A Short Course in the Secret War*. Lanham, MD: Madison Books.

Ferrell, Robert H. 1957. *American Diplomacy in the Great Depression: Hoover-Stimson Foreign Policy, 1929–1933*. New Haven, CT: Yale University Press.

Ficher M. 2011. "Social Network Analysis and Qualitative Comparative Analysis: Their Mutual Benefit for the Explanation of Policy Network Structures." *Methodological Innovations* 6(2): 27–51.

Filson, Darren, and Suzanne Werner. 2004. "Bargaining and Fighting: The Impact of Regime Type on War Onset, Duration, and Outcomes." *American Journal of Political Science* 48(2): 296–313.

Finnemore, Martha and Kathryn Sikkink. 1998. "International Norm Dynamics and Political Change." *International Organization* 52(4): 887–917.

First, Ruth. 1970. *The Barrel of a Gun: Political Power in Africa and the Coup D'état*. London: Allen Lane.

Flores, Alejandro Quiroz. 2012. "A Competing Risks Model of War Termination." *International Studies Quarterly* 56: 809–819.

Flynn, Mollie. 2014. "Peace Profile: Salvador Allende." *Peace Review* 26(2): 281–286.

Foran, John. 1991. "The Strengths and Weaknesses of Iran's Populist Alliance: A Class Analysis of the Constitutional Revolution of 1905–1911." *Theory and Society* 20(6): 795–823.

Foreign Relations of the United States, 1952–1954, IRAN, 1951–1954, Vol. X. The question of military and economic assistance to Iran; interest of the United States in the settlement of the Anglo-Iranian oil dispute, 1952–1954 (Documents 139–508).

Forsythe, David P. 1992. "Democracy, War, and Covert Action." *Journal of Peace Research* 29(4): 385–395.

Frantz Erica and Natasha Ezrow. 2011. *The Politics of Dictatorship: Institutions and Outcomes in Authoritarian Regimes*. Boulder, CO: Lynne Rienner.
Fraser, Andrew. 2005. "Architecture of a Broken Dream: The CIA and Guatemala, 1952–54." *Intelligence and National Security* 20(3): 486–508.
Freedman, Lawrence. 2001. *Kennedy's Wars: Berlin, Cuba, Laos, and Vietnam*. New York: Oxford University Press.
Freedom House. 1972–2006. *Freedom in the World*. Lanham, MD: Rowman & Littlefield.
Frederick, William H. and Robert L. Worden, editors. 1993. *Indonesia: A Country Study*. Washington, DC: GPO for the Library of Congress.
Friedman, Thomas. 1993. "U.S. Vision of Foreign Policy Reversed." *New York Times*, September 22, 1993.
Friedrich C. and Z. Brzezinski. 1956. *Totalitarian Dictatorship and Autocracy*. Cambridge, MA: Harvard University Press.
Fukuyama, Francis. 1991. "Liberal Democracy as a Global Phenomenon." *Political Science and Politics* 24(4) (December): 659–664.
Gaddis, John Lewis. 1989. "Looking Back: The Long Peace." *The Wilson Quarterly* 13(1): 42–65.
_____. 1992–1993. "International Relations Theory and the End of the Cold War." *International Security* 17(3): 5–58.
Galetovic, Alexander and Ricardo Sanhueza. 2000. "Citizens, Autocrats, and Plotters: A Model and New Evidence on Coups D'Etat." *Economics and Politics* 12(2): 183–204.
Gambone, Michael. 2014. *Small Wars: Low-Intensity Threats and the American Response since Vietnam*. Knoxville: University of Tennessee Press.
Garner, Robert J. 1990. *Ethical Guidelines for Military Covert Operations*. Carlisle Barracks, PA: US Army War College.
Garthoff, Raymond. 1989. *Reflections on the Cuban Missile Crisis*. Washington, DC: The Brookings Institution.
Gartzke, Erik. 2007. "The Capitalist Peace." *American Journal of Political Science* 51(1) (January): 166–191.
Gasiorowski, Mark J. and Malcolm Byrne. 2004. *Mohammed Mossadegh and the 1953 Coup in Iran*. Syracuse, NY: Syracuse University Press.
Gates, Scott, Havard Hegre, Mark R. Jones, and Håvard Strand. 2006. "Institutional Inconsistency and Political Instability: Polity Duration, 1800–2000." *American Journal of Political Science* 50 (October): 893–908.
Geddes, Barbara. 1999. "What Do We Know about Democratization after Twenty Years." *Annual Review of Political Science* 2: 115–144.
Geddes, Barbara, Joseph Wright, and Erica Frantz. 2014. "Autocratic Breakdowns and Regime Transitions: A New Dataset." *Perspectives on Politics* 12(2) (June): 313–331.
Gelpi, Christopher and Joseph M. Grieco. 2001. "Attracting Trouble: Democracy, Leadership Tenure, and the Targeting of Militarized Challenges, 1918–1992." *The Journal of Conflict Resolution* 45(6) (December): 794–817.

George, Alexander and Andrew Bennett. 2005. *Case Studies and Theory Development in the Social Sciences*. Boston, MA: MIT Press.

Gerard, Emmanual and Bruce Kuklick. 2015. *Death in the Congo: Murdering Patrice Lumumba*. Boston: Harvard University Press.

Getchell, Michelle Denise. 2015. "Revisiting the 1954 Coup in Guatemala." *Journal of Cold War Studies* 17(2): 73–102.

Gettleman, Jeffrey. 2009. "A Wound in the Heart of Africa." *New York Times Book Review,* April 5, 2009.

Gilpin, Robert. 1981. *War and Change in International Politics.* Cambridge, UK: Cambridge University Press.

Gisselquist, Rachel M. 2014. "Paired Comparison and Theory Development: Considerations for Case Selection." *Political Science and Politics* 47(2) (April): 477–484.

Glaser, Charles L. 1992. "Political Consequences of Military Strategy: Expanding and Refining the Spiral and Deterrence Models." *World Politics* 44(4) (July): 497–538.

_____. 2003. "Structural Realism in a More Complex World." *Review of International Studies* 29(3) (July): 403–414.

Gleditsch, Nils Petter, Peter Wallensteen, Mikael Eriksson, Margareta Sollenberg, and Håvard Strand. 2002. "Armed Conflict 1946–2001: A New Dataset." *Journal of Peace Research* 39(5): 615–37.

Gleditsch, H. E. Kristian and Giacomo Chiozza. 2009. "Introducing Archigos: A Data Set of Political Leaders." *Journal of Peace Research* 46(2): 269–83.

Gleijeses, Piero. 1995. "Ships in the Night: The CIA, the White House and the Bay of Pigs." *Journal of Latin American Studies* 27: 1–42.

Goemans, Henk E. 2000a. "Fighting for Survival: The Fate of Leaders and the Duration of War." *Journal of Conflict Resolution* 44(5): 555–579.

_____. 2000b. *War and Punishment: The Causes of War Termination and the First World War.* Princeton, NJ: Princeton University Press.

_____. 2008. "Which Way Out? The Manner and Consequences of Losing Office." *Journal of Conflict Resolution* 53(6): 771–794.

Goemans Henk, K. S. Gleditsch, and G. Chiozza. 2009. "Introducing Archigos: A Dataset of Political Leaders." *Journal of Peace Research* 46(2): 269–283.

Goehring, Jeannette. 2007. *Nations in Transit 2007: Democratization from Central Europe to Eurasia*. Washington, DC: Freedom House.

Godson, Roy. "Covert Action: An Introduction." In *Intelligence Requirements for the 1980s: Covert Action*, edited by Roy Godson. Washington, DC: National Strategy Information Center, Inc., 1985.

Godson, Roy. 2008. *Dirty Tricks or Trump Cards: U.S. Covert Action and Counterintelligence*. London: Transaction Publishers.

Goemans, H. E. 2000. "Fighting for Survival: The Fate of Leaders and the Duration of War." *The Journal of Conflict Resolution* 44(5) (October): 555–579.

Goldman, Francisco. 2007. *The Art of Political Murder*. New York: Grove Press.

Goldstone, Jack A., Robert H. Bates, David L. Epstein, Ted Robert Gurr, Michael B. Lustik, Monty G. Marshall, Jay Ulfelder, and Mark Woodward. 2010. "A Global Model for Forecasting Political Instability." *American Journal of Political Science* 54(1): 190–208.

Goldstone, Jack A. 1991. *Revolution and Rebellion in the Early Modern World*. Berkeley, CA: University of California Press.

Goodwin, Jeff. 2001. *No Other Way Out: States and Revolutionary Movements, 1945–1991*. Cambridge, UK: Cambridge University Press.

Grader, Sheila. 1988. "The English School of International Relations: Evidence and Evaluation." *Review of International Studies* 14(1) (January): 29–44.

Granville, Johanna. 2005. "Caught with Jam on Our Fingers: Radio Free Europe and the Hungarian Revolution of 1956." *Diplomatic History* 29(5): 811.

Greenberg, Harold M. "The Doolittle Commission of 1954." *Intelligence and National Security* 20(4) (2005): 687–694.

Gurr, Ted. 1970. *Why Men Rebel*. Princeton, NJ: Princeton University Press.

_____. 1994. "Peoples Against States: Ethnopolitical Conflict and the Changing World System: 1994 Presidential Address," *International Studies Quarterly* 38(3) (September): 347–377.

_____. 2000. "Ethnic Warfare on the Wane." *Foreign Affairs* 79 (May/June): 52–64.

Gustafson, Kristian. 2007. *Hostile Intent: U.S. Covert Operations in Chile, 1964–1974*. Washington, DC: Potomac Books.

Gutteridge, William F. 1962. *Armed Forces in New States*. London, UK: Oxford University Press.

Haber, Stephen. 2006. "Authoritarian Government." In *The Oxford Handbook of Political Economy*, edited by Barry R. Weingast and Donald A. Wittman, 693–707. Oxford, UK: Oxford University Press.

Haggard, Stephan, and Beth A. Simmons. 1987. "Theories of International Regimes." *International Organization* 41(3): 491–517.

Haines, Gerald K. 1998/1999. "The Pike Committee Investigations and the CIA." *Studies in Intelligence* 42: 81–92.

Hancock, Larry, and Stuart Wexler. 2014. *Shadow Warfare: The History of America's Undeclared Wars*. Berkeley, CA: Counterpoint.

Hansen, Morten T. 2013. "How John F. Kennedy Changed Decision Making for Us All." *Harvard Business Review*. https://hbr.org/2013/11/how-johnf-kennedy-changed-decision-making.

Harbom, Lotta, Stina Högbladh, and Peter Wallensteen. 2006. "Armed Conflict and Peace Agreements," *Journal of Peace Research* 43(5): 617–631.

Harbom, Lotta, Erik Melander, and Peter Wallensteen. 2008. "Dyadic Dimensions of Armed Conflict, 1946–2007." *Journal of Peace Research* 45(5): 697–710.

Harbom, Lotta and Peter Wallensteen. 2005. "Armed Conflict and Its International Dimensions, 1946–2004." *Journal of Peace Research* 42(5): 623–635.

Haslam, J. 2002. *No Virtue Like Necessity: Realist Thought in International Relations Since Machiavelli*. New Haven, CT: Yale University Press.

Hawkins, Jack. 1961. Record of Paramilitary Action against the Castro Government of Cuba 17 March 1960–May 1961. Washington, DC: National Security Archive, 5 May 1961. http://nsarchive.gwu.edu/latin_america/cuba/hawkins/. 22 Nov. 2016.

Henderson, William and Wesley R. Fishel. 1966. "The Foreign Policy of Ngo Dinh Diem." *Vietnam Perspectives* 2(1) (August): 3–30.

Henning, Tamm. 2016. "The Origins of Transnational Alliances: Rulers, Rebels, and Political Survival in the Congo Wars." *International Security* 41(1) (Summer): 147–181.

Henriksen, Thomas H. 2000. "Covert Operations, Now More than Ever." *Orbis* 44(1): 145.

Hermann, Margaret. 1980. "Explaining Foreign Policy Behavior Using the Personal Characteristics of Political Leaders." *International Studies Quarterly* 24(1) (March): 7–46.

Herring, George C. 2001. *America's Longest War: The United States and Vietnam, 1950–1975* (4th ed.), New York: McGraw-Hill.

Hersh, Seymour. 1974. "Huge CIA Operation Reported in U.S. Against Antiwar Forces, Other Dissidents in Nixon Years." *New York Times*, December 22, 1974.

Hibbs Jr., D. A. 1977. "Political Parties and Macroeconomic Policy." *American Political Science Review* 71(4): 1467–1487.

Higgott, Richard and Finn Fuglestad. 1975. "The 1974 Coup d'e tat in Niger: Towards an Explanation." *Journal of Modern African Studies* 13(3): 383–398.

Hilsman, Roger. 1995. "Does the CIA Still Have a Role?" *Foreign Affairs* 74(5) (September/October).

Hinckle, Warren and William W. Turner. 1992. *Deadly Secrets: The CIA–Mafia War against Castro and the Assassination of JFK*. New York: Thunder's Mouth Press.

Hobson, John M. 2000. *The State and International Relations*. Cambridge: Cambridge University Press.

Hoffman, Matthew. 1987. "Critical Theory and the Inter-Paradigm Debate." *Millennium: Journal of International Studies* 16(2): 231–249.

Hoffman, Stanley H. 1959. "International Relations: The Long Road to Theory." *World Politics* 11(3): 346–377.

Hopf, Ted. 1998. "The Promise of Constructivism in International Relations Theory." *International Security* 23(1) (Summer): 171–200.

Horowitz, Donald. 1990. "Comparing Democratic Systems," *Journal of Democracy* 1(4) (Fall): 73–79.

Horowitz, Donald. 2003. "Electoral Systems: A Primer for Decision-Makers." *Journal of Democracy* 14(4) (January): 115–127.

House, R. J. and J. M. Howell. 1992. "Personality and Charismatic Leadership." *Leadership Quarterly* 3: 81–108.

Hove, Mark T. 2007. "The Arbenz Factor: Salvador Allende, U.S.-Chilean Relations, and the 1954 U.S. Intervention in Guatemala." *Diplomatic History* 31(4): 623–663.

Huntington, Samuel P. 1968. *Political Order in Changing Societies*. New Haven, CT: Yale University Press.

_____. 1985. *Soldier and the State: The Theory and Politics of Civil Military Relations*. Cambridge, MA: Harvard University Press.

_____. 1991. *The Third Wave: Democratization in the Late Twentieth Century*. Norman, OK: University of Oklahoma Press.

_____. 1993. "The Clash of Civilizations?" *Foreign Affairs* 72(3) (Summer): 22–49.

_____. 1996. "Democracy for the Long Haul." *Journal of Democracy* 7(2): 3–13.

_____. 2000. "Try Again: A Reply to Russett, Oneal & Cox." *Journal of Peace Research* 37(5) (September): 609–610.

Hurrell, Andrew. 2001. "Keeping History, Law and Political Philosophy Firmly within the English School." *Review of International Studies* 27(3) (July): 489–494.

Husain, Aiyaz. 2005. "Covert Action and US Cold War Strategy in Cuba, 1961–62." *Cold War History* 5(1): 23–53.

Ido, Oren. 1995. "The Subjectivity of the 'Democratic' Peace: Changing U.S. Perceptions of Imperial Germany." *International Security* 20(2) (Autumn): 147–184.

Immerman, Richard H. 1982. *The CIA in Guatemala: The Foreign Policy of Intervention*. Austin: University of Texas Press.

Inglehart, Ronald and Christian Welzel. 2012. *Modernization, Cultural Change, and Democracy: The Human Development Sequence*. New York: Cambridge University Press.

Iqbal, Zaryab and Christopher Zorn. 2006. "Sic Semper Tyrannis? Power, Repression, and Assassinations since the Second World War." *Journal of Politics* 68(3): 489–501.

_____. 2008. "The Political Consequences of Assassination." *Journal of Conflict Resolution* 52(3): 385–400.

Isenberg, David. 1989. *The Pitfalls of U.S. Covert Action*. Washington, DC: Cato Institute Policy Analysis.

Ives, Christopher K. 2007. *U.S. Special Forces and Counterinsurgency in Vietnam: Military Innovation and Institutional Failure, 1961–1963*. London, UK: Routledge.

Jackman, Robert W. 1978. "The Predictability of Coups d'état: a Model with African Data." *American Political Science Review* 72: 1262–75.

Jaggers, Keith and Ted Robert Gurr. 1995. "Tracking Democracy's Third Wave with the Polity III Data." *Journal of Peace Research* 32(4) (November): 469–482.

_____. 1996. *POLITY III: Regime Change and Political Authority, 1800–1994*. ICPSR 6695. Ann Arbor, MI: Inter-university Consortium for Political and Social Research.

Janowitz, Morris. 1964. *The Military in the Political Development of New Nations*. Chicago: University of Chicago Press.

JCS/Joint Chiefs of Staff. 1961a. JCSM-57-61, Memorandum From the Joint Chiefs of Staff to Secretary of Defense McNamara – Military Evaluation of the Cuban Plan. Foreign Relations of the United States, 1961–1963, Volume X, Cuba, January 1961–September 1962, Document 35. Washington, DC: Office of the Historian, U.S. Department of State, 3 Feb. 1961. https://history.state.gov/historicaldocuments/frus1961-63v10/d35. 22 Nov. 2016.

_____. 1961b. JCSM-166-61, Memorandum From the Joint Chiefs of Staff to Secretary of Defense McNamara – Evaluation of the Military Aspects of Alternate Concepts, CIA Para-Military Plan, Cuba (S). Foreign Relations of the United States, 1961–1963, Volume X, Cuba, January 1961–September 1962, Document 62. Washington, DC: Office of the Historian, U.S. Department of State, 15 Mar. 1961. https://history.state.gov/historicaldocuments/frus196163v10/d62. 22 Nov. 2016.

_____. 1962. Memorandum for the Secretary of Defense re: Justification for Military Intervention in Cuba, 13th March 1962. http://www.gwu.edu/~nsarchiv/news/20010430/doc1.pdf

Jenkins W. O. 1947. "A Review of Leadership Studies with Particular Reference to Military Problems." *Psychological Bulletin* 44, 54–79.

Jervis, Robert. 1978. "Cooperation under the Security Dilemma." *World Politics* 30(2): 167–214.

_____. 1988. "War and Misconception." *The Journal of Interdisciplinary History* 18(4): 675–700.

_____. 1997. *System Effects: Complexity in Political Life and Social Life*. Princeton, NJ: Princeton University Press.

John F. Kennedy Presidential Library (JFKL). Papers of the Select Committee to Study Governmental Operations with Respect to Intelligence Activities of the United States Senate, Ninety-Fourth Congress, First Session (Church Committee), Box 1, Cuba, 1961–1962.

_____. National Security Files, Meetings and Memoranda Series, Box 313, NSC Meetings, 1961.

_____. Papers of Robert F. Kennedy, Attorney General Papers, Handwritten Notes.

_____. Personal Papers of Richard N. Goodwin.

_____. President's Office Files.

Johnson, Loch. 1980. "The CIA: Controlling the Quiet Option." *Foreign Policy* 39 (Summer): 143–52.

_____. 1987. *America's Secret Power: The CIA in a Democratic Society*. New York: Oxford University Press.

_____. 1989. "Covert Action and Accountability: Decision-Making for America's Secret Foreign Policy." *International Studies Quarterly* 33(1): 81–109.

_____. 1991. *America's Secret Power: The CIA in a Democratic Society.* New York: Oxford University Press.

_____. 1992. "On Drawing a Bright Line for Covert Operation," *American Journal of International Law* 86(2) (April): 286.

_____. 1998. *Secret Agencies: U.S. Intelligence in a Hostile World*. New Haven, CT: Yale University Press.

_____. 2011. *Intelligence: Critical Concepts in Military, Strategic & Security Studies, Vol II: Covert Action, the Aggressive Arm of National Security Intelligence*. New York: Routledge.

_____. 2012. "Intelligence Analysis and Planning for Paramilitary Operations." *Journal of National Security Law & Policy* 5: 481–506.

_____. 2014. "The Myths of Covert Action." *Virginia Policy Review* 7 (Winter): 52–64.

_____. 2015. *A Season of Inquiry Revisited: The Church Committee Confronts America's Spy Agencies.* Lawrence: University Press of Kansas.

_____. 2017. *National Security Intelligence: Secret Operations in Defense of the Democracies.* 2nd Edition. Cambridge: Polity Press.

_____. 2018. *Spy Watching: Intelligence Accountability in the United States*. New York: Oxford University Press.

Johnson, Thomas, Robert Slater, and Pat McGowan. 1984. "Explaining African Military Coups d'état, 1960–1982." *American Political Science Review* 78: 622–40.

Jones, Benjamin F., and Benjamin A. Olken. 2005. "Do Leaders Matter? National Leadership and Growth Since World War II." *Quarterly Journal of Economics* 120(3): 835–864.

_____. 2007. "Hit or Miss? The Effect of Assassinations on Institutions and War." *American Economic Journal* 1(2): 55–87.

Jones, Clive. 2004. "Among Ministers, Mavericks, and Mandarins': Britain, Covert Action, and the Yemen Civil War, 1962–1964." *Middle East Studies* 40(1) (January): 99–126.

Jones, Howard. 2003. *Death of a Generation: How the Assassinations of Diem and JFK Prolonged the Vietnam War*. New York: Oxford University Press.

Jones, Roy. 1981. "The English School of International Relations: A Case for Closure." *Review of International Studies* 7(1) (January): 1–13.

Judge, Timothy A. and Joyce E. Bono. 2000. "Five-Factor Model of Personality and Transformational Leadership." *Journal of Applied Psychology* 85(5): 751–765.

_____. 2002. "Five-Factor Model of Personality and Transformational Leadership." *Journal of Applied Psychology* 85: 751–765.

Juliana Lumumba, interview by Kinshasa, *Publisher*, Network, November 18, 2006.

Kaplan, Roger, ed. 1996. "Cuba" in *Freedom in the World: The Annual Survey of Political Rights and Civil Liberties 1995–1996*. New York: Freedom House, 199.

Karl, Terry Lynn. 2000. "Economic Inequality and Democratic Instability." *Journal of Democracy* 11(1) (January): 149–156.

Karnow, Stanley. 1997. *Vietnam: A History* (2nd ed.). New York: Penguin Books.

Kaufmann, D., A. Kraay, and M. Mastruzzi 2011. "The Worldwide Governance Indicators: Methodology and Analytical Issues." *Hague Journal on the Rule of Law* 3(2): 220–246.

Kennan, George F. (writing as "X"). 1947. "Sources of Soviet Conduct." *Foreign Affairs* 25(4).

Keohane, Robert. 1984. *After Hegemony*. Princeton, NJ: Princeton University Press.

_____. 1986. *Neorealism and Its Critics*. New York: Columbia University Press.

_____. 2005. "Abuse of Power: Assessing Accountability in World Politics." *Harvard International Review* 27(2): 48–53.

Keohane, Robert and Joseph Nye. 2000. *Power and Interdependence*. New York: Longman.

Kibbe, Jennifer D. 2007. "Covert action and the Pentagon." *Intelligence and National Security* 22(1): 57–74.

Kim, Jaechun. 2002. "U.S. Covert Action in Indonesia in the 1960s: Assessing the Motives and Consequences." *Journal of International and Area Studies* 9(2): (December): 63.

King, Gary. 1989. *Unifying Political Methodology: The Likelihood Theory of Statistical Inference*. New York: Cambridge University Press.

King, Gary and Langche Zeng. 2001a. "Logistic Regression in Rare Events Data." *Political Analysis* 9 (Spring): 137–63.

_____. 2001b. "Improving Forecasts of State Failure." *World Politics* 53 (July): 623–58.

Kinzer, Stephen. 2006. *Overthrow: America's Century of Regime Change from Hawaii to Iraq*. New York: Time Books.

Kinzer, Stephen. 2008. *All the Shah's Men*. Hoboken, NJ: John Wiley & Sons.

Klein, J. and M. Moeschberger. 1997. *Survival Analysis: Techniques for Censored and Truncated Data*. New York: Springer.

Knaus, John Kenneth. 2003. "Official Policies and Covert Programs: The U.S. State Department, the CIA, and the Tibetan Resistance." *Journal of Cold War Studies* 5(3) (Summer): 54–79.

Knott, Stephen F. 1996. *Secret and Sanctioned: Covert Operations and the American Presidency*. New York: Oxford University Press.

Kohli, Atul. 2004. *State-Directed Development: Political Power and Industrialization in the Global Periphery*. New York: Cambridge University Press.

Kornbluh, Peter, ed. 1998. *The Bay of Pigs Declassified: The Secret CIA Report on the Invasion of Cuba*. New York: The New Press.

Krasner, Stephen. 1979. "State Power and the Structure of International Trade." *World Politics* 28(3): 317–347.

Krepinevich, Andrew F. 1986. *The Army and Vietnam*. Baltimore, MD: Johns Hopkins University Press.

Kreutz, Joakim. 2010. "How and When Armed Conflicts End: Introducing the UCDP Conflict Termination Dataset." *Journal of Peace Research* 47(2) (March): 243–250.

Kuklick, Bruce. 2014. "Killing Lumumba." *Proceedings of the American Philosophical Society* 158(2) (June): 144–152.

Kupchan, Charles A. and Clifford A. Kupchan. "The Promise of Collective Security." *International Security* 20(1) (Summer 1995): 52–61.

Kvale, S. 1989. *Issues of Validity in Qualitative Research*. Lund, Sweden: Chartwell Bratt.

Kydd, Andrew H. 2005. *Trust and Mistrust in International Relations*. Princeton, NJ: Princeton University Press.

Laurie, Clayton. 2016. "CIA and the Wars in Southeast Asia, 1947–1975." *Journal of the American Intelligence Professional* (August).

Lawson, Stephanie. 1993. "Conceptual Issues in the Comparative Study of Regime Change and Democratization." *Comparative Politics* 25(2) (January): 183–205.

Leary, William M. 1984. *The Central Intelligence Agency: History and Documents*. Tuscaloosa: The University of Alabama Press.

Lefever, Ernest W. 1970. *Spear and Scepter: Army, Police, and Politics in Tropical Africa*. Washington, DC: The Brookings Institution.

Legge, J. D. 1972. *Sukarno: A Political Biography*. New York: Penguin Press.

Lentz, Harris M., III. 1988. *Assassinations and Executions: An Encyclopedia of Political Violence, 1865 through 1986*. Jefferson, NC: McFarland & Company.

_____. 1994. *Heads of States and Governments: A Worldwide Encyclopedia of over 2,300 Leaders, 1945 through 1992*. Jefferson, NC: McFarland & Company.

_____. 1999. *Encyclopedia of Heads of states and Governments, 1900 through 1945*. Jefferson, NC: McFarland & Company.

_____. 2002. *Assassinations and Executions: An Encyclopedia of Political Violence, 1900 through 2000*. Jefferson, NC: McFarland & Company.

LeoGrande, William M. 1982. "Cuba Policy Recycled." *Foreign Policy* 46 (Spring): 105–119.

Levitsky, Steven and Lucan Way. 2002. "The Rise of Competitive Authoritarianism," *Journal of Democracy* 13(2) (April): 51–65.

_____. 2010. *Competitive Authoritarianism: Hybrid Regimes After the Cold War*. New York: Cambridge University Press.

Levy, Jack S. 1994. "Learning and Foreign Policy: Sweeping a Conceptual Minefield." *International Organization* 48(2) (Spring): 279–312.

Lieberson, Stanley. 2004. "Comments on the Use and Utility of QCA." *Qualitative Methods* 2(2): 13–14.

Lijphart, Arendt. 1991. "Majority Rule in Theory and Practice: The Tenacity of a Flawed Paradigm." *International Social Science Journal* 129 (August): 483–493.

_____. 1999. *Patterns of Democracy: Government Forms and Performance in Thirty-Six Countries*. New Haven, CT: Yale University Press.

_____. 2004. "Constitutional Design for Divided Societies." *Journal of Democracy* 15(2) (April): 96–109.

Linklater, Andrew and Suganami, H. 2006. *The English School of International Relations: A Contemporary Reassessment*. Cambridge, UK: Cambridge University Press.

Linz, Juan. 1990. "Presidents vs. Parliaments: The Virtues of Parliamentarism." *Journal of Democracy* 1(4): 84–91.

_____. 2000. *Totalitarian and Authoritarian Regimes*. Boulder, CO: Lynne Reiner.

Linz, Juan and A. Stepan, eds., 1978. *The Breakdown of Democratic Regimes*. Johns Hopkins University Press.

Lipset, Seymour Martin. 1959. "Some Social Requisites for Democracy: Economic Development and Political Legitimacy." *American Political Science Review* 53(1): 69–105.

Little, Richard. 2003. "The English School vs. American Realism: A Meeting of Minds or Divided by a Common Language?" *Review of International Studies* 29(3) (July): 443–460.

Lo, Nigel, Barry Hashimoto, and Dan Reiter. 2008. "Ensuring Peace: Foreign-Imposed Regime Change and Postwar Peace Duration: 1914–2001." *International Organization* 62(4) (Fall): 717–736.

Lowenthal, Mark M. 2009. *Intelligence: From Secrets to Policy*. Washington, DC: CQ Press.

Lucas, Scott. 1996. "Campaigns of Truth: The Psychological Strategy Board and American Ideology, 1951–1953." *International History Review* 18(2) (May): 279–302.

Luckham, Robin. 1994. "The Military, Militarisation and Democratisation in Africa: A Survey of Literature and Issues." *African Studies Review* 37(2): 13–75.

Luttwak, Edward. 1968. *Coup D'E'tat: A Practical Handbook*. Cambridge, MA: Harvard University Press.

Maeda, Ko. 2006. "Duration of Party Control in Parliamentary and Presidential Governments: A Study of 65 Democracies, 1950 to 1998." *Comparative Political Studies* 39(3) (April): 352–374.

Maggetti, Martino and David Levi-Faur. 2013. "Dealing with Errors in QCA." *Political Research Quarterly* 66(1) (March): 198–204.

Mahoney, James. 2004. "Comparative Historical Methodology." *Annual Review of Sociology* 30: 81–101.

Mahoney, James and Gary Goertz. 2006. "A Tale of Two Cultures: Contrasting Quantitative and Qualitative Research." *Political Analysis* 14(3): 227–249.

Mahoney, James and Rueschemeyer, Dietrich. 2003. "Comparative Historical Analysis: Achievements and Agendas." In *Comparative Historical Analysis in the Social Sciences*, edited by James Mahoney and Dietrich Rueschemeyer. Cambridge: Cambridge University Press.

Mansfield, Edward and J. Snyder. 1995. "Democratization and the Danger of War." *International Security* 20(1) (Summer): 5–38.

_____. 2002. "Incomplete Democratization and the Outbreak of Military Disputes." *International Studies Quarterly* 46(4) (December): 529–549.

_____. 2002. "Democratic Transitions, Institutional Strength, and War." *International Organization* 56(2) (Spring): 297–337.

_____. 2005. *Electing to Fight: Why Emerging Democracies Go to War*. Cambridge, MA: MIT Press.

Marshall, M. G. and K. Jaggers. 2002. "Polity IV Project: Political Regime Characteristics and Transitions, 1800–2002." http://www.systemicpeace.org/polity/polity4.htm

Martin, Guy. 2005. "Conflict in the Congo: Historical and Regional Perspectives." *African Studies Review* 48(1): 127–137.

Martin, James W. 2013. "Becoming Banana Cowboys: White-Collar Masculinity, the United Fruit Company and Tropical Empire in Early Twentieth-Century Latin American." *Gender & History* 25(2) (August): 317–338.

Marx, Axel, Benoit Rihoux, and Charles Ragin. 2014. "The Origins, Development, and Application of Qualitative Comparative Analysis: The First 25 Years." *European Political Science Review* 6(1) (February): 115–142.

Mastanduno, Michael, David Lake, and John Ikenberry. 1989. "Toward a Realist Theory of State Action." *International Studies Quarterly* 33(4): 457–74.

McGeorge, Bundy. 1964. "Presidential Election in Chile," memorandum to Lyndon B. Johnson, May 13, 1964, NSF, Country File, vol. 1, 1/64-8/64, Memos to the President, McGeorge Bundy, vol. 4, LBJPA.

McGillivray, Fiona and Allan C. Stam. 2004. "Political Institutions, Coercive Diplomacy, and the Duration of Economic Sanctions." *The Journal of Conflict Resolution* 48(2) (April): 154–172.

McGregor, Katherine. 2016. "Cold War Scripts: Comparing Remembrance of the Malayan Emergency and the 1965 Violence in Indonesia." *South East Asia Research* 24(2) (June): 242–260.

McMahon, Robert. 1986. "Eisenhower and Third World Nationalism: A Critique of the Revisionists." *Political Science Quarterly* 101: 453–53.

McNaughton, Colm. 2015. "Reading the Mass Violence in Indonesia 1965–1966 as a Form of Primitive Accumulation." *Inter-Asia Cultural Studies* 16(2): 292–305.

McRae, Robert R. 1987. "Creativity, Divergent Thinking, and Openness to Experience." *Journal of Personality and Social Psychology*, 52: 1258–1265.

Meade, Teresa. 2012. "Holding the Junta Accountable: Chile's 'Sitios de Memoria' and the History of Torture, Disappearance, and Death." *Radical History Review* 79 (Winter): 123–139.

Meadows, Donella. *Thinking in Systems: A Primer.* White River Junction, VT: Chelsea Green Publishing Company, 2008.

Mearsheimer, John J. 1979. *The Tragedy of Great Power Politics*. New York: W. W. Norton & Company.

_____. 1990. "Back to the Future: Instability in Europe after the Cold War." *International Security* 15(1): 5–56.

_____. 1994/95. "The False Promise of International Institutions." *International Security* 19(3): 5–49.

_____. 1995. "A Realist Reply." *International Security* 20(1): 82–93.

_____. 2005. "E. H. Carr versus Idealism: The Battle Rages On." *International Relations* 19: 139–52.

Melley, Timothy. 2012. *Covert Sphere: Secrecy, Fiction, and the National Security State*. NY: Cornell University Press.

Memorandum for the Director of Central Intelligence. 1962. Meeting with the Attorney General of the United States concerning Cuba, Richard Helms, 19th January 1962 http://www.gwu.edu/~nsarchiv/nsa/cuba_mis_cri/620119%20Meeting%20 wit h%20the%20Attorney%20Gen.pdf

Memorandum for the Chief of Operations. 1962. Cuba Project, and attached Cover and Deception Plans, 19th February 1962. http://www.gwu.edu/~nsarchiv/nsa/cuba_mis_cri/620219%20Memo%20for %20the%20Chief%20of%20Opspdf

Memorandum for the record. 1962. Meeting with the President, Ed Lansdale, 16th March 1962. http://www.maryferrell.org/mffweb/archive/viewer/showDoc .do?absPageId=4 9622

Mercer, Jonathan. 1996. *Reputation and International Politics.* Ithaca, NY: Cornell University Press.

Milani, Abbas. 2015. "Iran's Paradoxical Regime." *Journal of Democracy* 26(2): 52–60.

Mingst, Karen A. and Ivan M. Arreguin-Toft. 2011. *Essentials of International Relations*, 5th ed. New York: W. W. Norton & Company.

Mirsky, Jonathan. 2013. "Tibet: The CIA's Cancelled War." *The New York Review of Books*, April 9, 2013.

Moaddel, Mansoor. 1996. "The Social Bases and Discursive Context of the Rise of Islamic Fundamentalism: The Cases of Iran and Syria." *Sociological Inquiry* 66(3) (July): 330–355.

Moe, Terry M. and Michael Caldwell. 1994. "The Institutional Foundations of Democratic Government: A Comparison of Presidential and Parliamentary Systems."

Journal of Institutional and Theoretical Economics (JITE) / Zeitschrift für diegesamte Staatswissenschaft, Symposium on The New Institutional Economics Bounded Rationality and the Analysis of State and Society 150(1) (March): 171–195.

Moore, Barrington. 1966. *Social Origins of Dictatorship and Democracy: Lord and Peasant in the Making of the Modern World,* Boston: Beacon Press.

Moravcsik, Andrew. 1997. "Taking Preferences Seriously: A Liberal Theory of International Politics." *International Organization* 51(4) (Autumn): 513–553.

Morgenthau, Hans J. 1948. *Politics Among Nations: The Struggle for Power and Peace*. New York: Knopf.

Morse, Janice, Michael Barrett, Maria Mayan, Karin Olson, Jude Spiers. 2002. "Verification Strategies for Establishing Reliability and Validity in Qualitative Research." *International Journal of Qualitative Methods.* 1(2): 13–22.

National Archives, RG 59, Central Files 1950–1954, 788.00/2–1652.

The National Archives (TNA) (Public Records Office) CAB 148/19 OPD (65) 25, 26 January 1965; National Intelligence Memorandum NIE 54/55-65, 1 July 1965, FRUS, Indonesia 1964–68, vol. 26, 270–71.

National Security Action Memos. Memos 55, 56 and 57, 28th June 1961. http://www.jfklibrary.org/ (images) http://www.ratical.org/ratville/JFK/USO/appE.html (text)

National Security Strategy of the United States (Washington, DC: 2017).

Nordlinger Eric. 1977. *Soldiers in Politics: Military Coups and Governments*. Englewood Cliffs, NJ: Prentice-Hall.

Nutter, John J. 2000. *The CIA's Black Ops: Covert Action, Foreign Policy, and Democracy.* New York: Prometheus Books.

Nzongola-Ntalaja. 2002. *The Congo from Leopold to Kabila: A People's History.* London: Zed Books.

O'Brien, Keith. "Interfering with Civil Society: CIA and KGB Covert Political Action During the Cold War." *International Journal of Intelligence and Counterintelligence* 8(4) (1995): 431–456.

O'Kane, Rosemary. 1981. "A Probabilistic Approach to the Causes of Coups d'Etat." *British Journal of Political Science* 11: 287–308.

O'Rourke, Lindsey. 2013. "Secrecy and Security: U.S.-Orchestrated Regime Change during the Cold War." PhD diss., University of Chicago.

Olson, Mancur. 1993. "Dictatorship, Democracy, and Development." *American Political Science Review* 87 (September): 567–76.

Origins of the Insurgency in South Vietnam, 1954–1960 (http:/ / www.mtholyoke.edu/ acad/intrel/ pentagon/ pent14.htm), The Pentagon Papers (Gravel Edition), Volume 1, Chapter 5, (Boston: Beacon Press, 1971), Section 3, pp. 314–346; International Relations Department, Mount Holyoke College.

Ottaway. 2003. "The Challenges of Semi-Authoritarianism: An Introduction." *Democracy Challenged: The Rise of Semi-Authoritarianism.* Washington, DC: Carnegie Endowment for International Peace. http://www.carnegieendowment.org/files/DemChallenged_Intro.pdf

Papers of John F. Kennedy. Presidential Papers. President's Office Files. Countries. Congo: General, 1959–1960.

Paul Wimert. 1999. "Backyard," interview for the Washington University National Security Archive, *The Cold War,* CNN, February 21, 1999.

Pérez, Louis A., Jr. 2002. "Fear and Loathing of Fidel Castro: Sources of US Policy toward Cuba." *Journal of Latin American Studies* 34(2) (May): 227–254.

Perry, David. L. 1995. "Repugnant Philosophy: Ethics, Espionage, and Covert Action." *Journal of Conflict Studies* (Spring): 92–115.

Peterzell, Jay. 1984. *Reagan's Secret Wars.* Washington, DC: The Center for National Security Studies.

Pfeiffer, Jack B. 1979. Evolution of the CIA's Anti-Castro Policies, 1959–January 1961. Langley: Central Intelligence Agency Library, December 1979. https://www.cia.gov/library/readingroom/docs/bop-vol3.pdf.

_____. 1984. The Taylor Committee Investigation of the Bay of Pigs. Langley: Central Intelligence Agency Library, 9 November 1984. https://www.cia.gov/library/readingroom/docs/bop-vol4.pdf.

Pickering Jeffrey and Mark Peceny. 2006. "Forging Democracy at Gunpoint." *International Studies Quarterly* 50(3): 539–60.

Polity IV. 2010. "Political Regime Characteristics and Transitions, 1800–2010." http://www.systemicpeace.org/polity/polity4.htm.

Power, Margaret. 2009. "The U.S. Movement in Solidarity with Chile in the 1970s." *Latin American Perspectives* 36(6) (November): 46–66.

Poznansky, Michael. 2015. "Stasis or Decay? Reconciling Covert War and the Democratic Peace." *International Studies Quarterly* 59(4): 815–826.

Prados, John. 1996. *Presidents' Secret Wars: CIA and Pentagon Covert Operations from World War II through the Persian Gulf.* Chicago: Ivan R. Dee.

———. 2006. *Safe for Democracy: The Secret Wars of the CIA.* Chicago: Ivan R. Dee.

Prevost, Gary. 2007. "Fidel Castro and the Cuban Revolution." *Headwaters: The Faculty Journal of the College of Saint Benedict and Saint John's University* 24: 19–33.

Przeworski A. and H. Teune. 1968. *The Logic of Comparative Social Inquiry.* New York: Wiley.

Przeworksi, Adam and Fernando Limongi. 1973. "Modernization: Theories and Facts." *World Politics* 49 (August): 155–183.

Przeworski, Adam, Michael E. Alvarez, Jose Antonio Cheibub, and Fernando Limongi. 2000. *Democracy and Development.* Cambridge: Cambridge University Press.

Quinlivan, James T. 1999. "Coup-Proofing: Its Practice and Consequences in the Middle East." *International Security* 24(2): 131–165.

R Core Team 2018. *R: A Language and Environment for Statistical Computing.* R Foundation for Statistical Computing, Vienna, Austria. https://www.R-project.org/.

Rabkin, Rhoda. 1995. "Human Rights and Military Rule in Cuba," in *Cuban Communism*, 8th ed., edited by Irving Louis Horowitz. New Brunswick, NJ: Transaction Publishers, 671–672.

Ragin, Charles C. 1987. *The Comparative Method: Moving Beyond Qualitative and Quantitative Strategies.* Berkeley, CA: University of California Press.

———. 2000. *Fuzzy-Set Social Science*. Chicago, IL: University of Chicago Press.

———. 2008. *Redesigning Social Inquiry: Fuzzy Sets and Beyond.* Chicago, IL: University of Chicago Press.

Ranelagh, John. 1987. *The Agency: The Rise and Decline of the CIA.* New York: Simon and Schuster.

Ranstorp, Magnus. 1997. *Hizb'allah in Lebanon: The Politics of the Western Hostage Crisis*. New York: St. Martin's Press.

Regan, P. 2002. "Third Party Interventions and the Duration of Interstate Conflicts." *Journal of Conflict Resolution* 46: 55–73.

Regan, Patrick M. and Errol A. Henderson. 2002. "Democracy, Threats, and Political Repression in Developing Countries: Are Democracies Internally Less Violent?" *Third World Quarterly* 23(1): 119–136.

Reiter, Dan. 2003. "Exploring the Bargaining Model of War." *Perspectives on Politics* 1(1): 27–43.

Reiter, Dan and Allan C Stam. 2002. *Democracies at War*. Princeton, NJ: Princeton University Press.

Reus-Smit, Christian. 1997. "The Constitutional Structure of International Society and the Nature of Fundamental Institutions." *International Organization* 51(4) (Autumn): 555–589.

Report by the Department of Defense and the Joint Chiefs of Staff Representative. 1962. On the Caribbean Survey Group to the Joint Chiefs of Staff on Cuba Project, 9th March 1962 http://www.maryferrell.org/mffweb/archive/viewer/showDoc.do?absPageld=1 59978

Richards, Gordon. 1986. "Stabilization Crises and the Breakdown of Military Authoritarianism in Latin America." *Comparative Political Studies* 18 (January): 447–85.

Robarge, David. 2014. "CIA's Covert Operations in the Congo, 1960–1968: Insights from Newly Declassified Documents," *Studies in Intelligence* 58(3) (September): 1–9.

Robinson, Colin. 2012. "Army Reconstruction in the Democratic Republic of the Congo 2003–2009." *Small Wars & Insurgencies* 23(3): 474–499.

Roorda, Eric Paul. 1996. "Genocide Next Door: The Good Neighbor Policy, the Trujillo Regime, and the Haitian Massacre of 1937." *Diplomatic History* 20(3): 301–19.

———. 1998. *The Dictator Next Door: The Good Neighbor Policy and the Trujillo Regime in the Dominican Republic, 1930–1945*. Durham, NC: Duke University Press.

Roosevelt, Kermit. 1981. *Countercoup.* New York: McGraw-Hill.

Rosato, Sebastian. 2007. "Explaining the Democratic Peace." *American Political Science Review* 99(3) (September): 467–472.

Rose, Gideon. 1998. "Neoclassical Realism and Theories of Foreign Policy." *World Politics* 51(1): 144–172.

Rosen, Stephen. 2005. *War and Human Nature*. Princeton, NJ: Princeton University Press.

Rosenau, William and Austin Long. 2009. *The Phoenix Program and Contemporary Counterinsurgency*. RAND National Defense Research Institute. Santa Monica, CA: RAND Corporation.

Rositzke, Harry. 1988. *The CIA's Secret Operations: Espionage, Counterespionage, and Covert Action.* Boulder, CO: Westview Press.

Ross, Michael L. 2004. "What Do We Know about Natural Resources and Civil War?" *Journal of Peace Research* 41 (May): 337–56.

Rousseau, David. 2005. *Democracy and War: Institutions, Norms, and the Evolution of International Conflict.* Stanford, CA: Stanford University Press.

Rowold J. and M. Kersting. 2008. "The Assessment of Charismatic Leadership: Validity of a German Version of the Conger-Kanungo Scale (CKS)." *European Journal of Psychological Assessment* 24: 124–130.

Rudgers, David F. 2000. "The Origins of Covert Action." *Journal of Contemporary History* 35(2): 249–262.

Ruggie, John Gerard. 1983. "Continuity and Transformation in the World Polity: Toward a Neorealist Synthesis." *World Politics* 35(2): 261–285.

_____. 1993. "Territoriality and Beyond: Problematizing Modernity in International Relations." *International Organization* 47(1) (Winter): 139–174.

Sabatier, P. and D. Mazmanian. 1980. "The Implementation of Public Policy: A Framework of Analysis." *Policy Studies Journal* 8(4): 538–560.

Sager, F. and C. Andereggen. 2012. "Dealing with Complex Causality in Realist Synthesis: The Promise of Qualitative Comparative Analysis (QCA)." *American Journal of Evaluation* 33(1): 60–78.

Sale, Richard. 2009. *Clinton's Secret Wars: The Evolution of a Commander in Chief.* New York: St. Martin's Press.

Sambanis, Nicholas. 2004. "What is Civil War? Conceptual and Empirical Complexities of an Operational Definition." *The Journal of Conflict Resolution* 48(6): 814–58.

Sambanis, Nicholas. 2004. "Using Case Studies to Expand Economic Models of Civil War." *Perspectives on Politics* 2: 259–279.

Sarat, Austin. 2010. *Dissent in Dangerous Times*. Ann Arbor: University of Michigan Press.

Sarkees, Meredith Reid. 2000. "The Correlates of War Data on War: An Update to 1997." *Conflict Management and Peace Science* 18(1): 123–144.

Sarkees, Meredith Reid and Frank Wayman. 2010. *Resort to War: 1816–2007*. Washington, DC: CQ.

Sartori, Giovanni. 1987. *The Theory of Democracy Revisited.* Chatham, NJ: Chatham House.

Saunders, Richard M. 1985. "Military Force in the Foreign Policy of the Eisenhower Presidency." *Political Science Quarterly* 100(1) (Spring): 97–116.

Schaefer, B. 2013. "Indonesia and the World in 1965." *Journal of Sino-Western Communications* 5(2): 5–12.

Schelling, Thomas. 1960. *The Strategy of Conflict.* Cambridge, MA: Harvard University Press.

Schirmer, Jennifer. 1999. *The Guatemalan Military Project: A Violence Called Democracy*. Philadelphia: University of Pennsylvania Press.

Schlesinger, Stephen, and Stephen Kinzer. 2005. *Bitter Fruit: The Story of the American Coup in Guatemala*. Cambridge, MA: David Rockefeller Center for Latin American Studies.

Schmitter, Phillippe and T. Karl. 1991. "What Democracy is . . . and is Not." *Journal of Democracy* 2 (Summer): 75–88.

Schmitter, Phillippe. 1995. "The Consolidation of Political Democracies: Rhythms, Sequences, and Types." In *Transitions to Democracy*, edited by G. Pridham. Aldershot: Dartmouth University Press.

Schneider, Carsten Q. and Claudius Wagemann. 2010. "Standards of Good Practice in Qualitative Comparative Analysis (QCA) and Fuzzy-Sets." *Comparative Sociology* 9(3): 397–418.

_____. 2012. *Set-Theoretic Methods for the Social Sciences. A Guide to Qualitative Comparative Analysis*. Cambridge: Cambridge University Press.

Schneider, C. Q. and Ingo Rohlfing. 2013. "Combining QCA and Process Tracing in Set-Theoretic Multi-Method Research." *Sociological Methods & Research* 42(4): 559–597.

Schroeder, Paul. 1994. "Historical Reality vs. Neo-Realist Theory." *International Security* 19(1): 108–48.

Schultz, Kenneth A. 1999. "Do Democratic Institutions Constrain or Inform? Contrasting Two Institutional Perspectives on Democracy and War." *International Organization* 53(2) (Spring): 233–266.

Schultz, Richard H., Jr. 2000. "The Great Divide: Strategy and Covert Action in Vietnam." *Joint Force Quarterly* 23: 90–96.

Schweller, Randall. 2004. "Unanswered Threats: A Neoclassical Realist Theory of Underbalancing." *International Security* 29(2): 159–201.

Scott, Len. 2004. "Secret Intelligence, Covert Action and Clandestine Diplomacy," *Intelligence & National Security* 19, no. 2: 322–341.

Seawright, Jason. 2004. "Qualitative Comparative Analysis vis-à-vis Regression." *Qualitative Methods* 2(2): 14–17.

Sen, Amartya. 1999. "Democracy as a Universal Value." *Journal of Democracy* 10(3) (July): 3–17.

Sewell, Bevan. 2011. "The Pragmatic Face of the Covert Idealist: The Role of Allen Dulles in US Policy Discussions on Latin America, 1953–61." *Intelligence and National Security* 26(2–3): 269–290.

Shadish, W. R., Cook, T. D., and Campbell, D. T. 2002. *Experimental and Quasi-Experimental Designs for Generalized Causal Inference*. Boston, MA: Houghton Mifflin.

Shaw, Geoffrey W. 2015. *The Lost Mandate of Heaven: The American Betrayal of Ngo Dinh Diem, President of Vietnam*. San Francisco, CA: Ignatius Press.

Sigmund, Paul E. 1974. "The Invisible Blockade and the Overthrow of Allende." *Foreign Affairs* 52(2) (January): 322–340.

Sirin, Cigdem V. and Michael T. Koch. 2015. "Dictatorships and Death: Casual Sensitivity of Autocracies in Militarized Interstate Disputes." *International Studies Quarterly*, 59: 802–814.

Skaaning, S. 2011. "Assessing the Robustness of Crisp-set and Fuzzy-set QCA Results." *Sociological Methods and Research* 40(2): 391–408.

Skocpol, T. 1979. *States and Social Revolutions*. Cambridge, UK: Cambridge University Press.

Slotkin, R., 1972. *Regeneration through Violence: The Mythology of the American Frontier, 1600–1860*. Middletown, CT: Wesleyan University Press.

Slotkin, R., 1985. *The Fatal Environment: The Myth of the Frontier in the Age of Industrialization, 1800–1890*. New York: Atheneum.

Slotkin, R., 1992. *Gunfighter Nation: The Myth of the Frontier in Twentieth-Century America*. New York: Atheneum.

Snyder, Jack. 1991. *Myths of Empire*. Ithaca, NY: Cornell University Press.

Snyder, Richard, and James Mahoney. 1999. "The Missing Variable: Institutions and the Study of Regime Change." *Comparative Politics* 32 (October): 103–22.

Sorensen, Georg. 2007. *Democracy and Democratization in a Changing World: Process and Prospects in a Changing World*. Boulder, CO: Westview Press.

Stempel, John D. 2007. "Covert Action and Diplomacy." *International Journal of Intelligence and Counterintelligence* 20(1) (September): 122–135.

Sterling-Folker, Jennifer. 2013. *Making Sense of International Relations*. 2nd ed. Boulder, CO: Lynne Reinner Publishers.

Stiefler, Todd. 2004. "CIA's Leadership and Major Covert Operations: Rogue Elephants or Risk-Averse Bureaucrats?" *Intelligence & National Security* 19(4): 632–654.

Stogdill, R. M. 1948. "Personal Factors Associated with Leadership: A Survey of the Literature." *Journal of Psychology* 25: 35–71.

Sudduth, Jun Koga. 2016. "Coup-proofing and Civil War." In *The Oxford Research Encyclopedia of Politics*, edited by William R Thompson. Oxford: Oxford University Press.

_____. (forthcoming). "Strategic Logic of Elite Purges in Dictatorships." *Comparative Political Studies*.

Summary of Facts Investigation of CIA Involvement in Plans to Assassinate Foreign Leaders. 1975. Ann Arbor, MI: Richard B. Cheney Files at the Gerald R. Ford Presidential Library.

Svolik, Milan W. 2009. "Power Sharing and Leadership Dynamics in Authoritarian Regimes." *American Journal of Political Science* 53(2): 477–494.

_____. 2012. *The Politics of Authoritarian Rule*. New York, NY: Cambridge University Press.

_____. 2013. "Contracting on Violence: Authoritarian Repression and Military Intervention in Politics." *Journal of Conflict Resolution* 57(5): 765–794.

Taghipour, Jahangir. 2006. "Mossadegh and the People of Iran." *Medicine, Conflict and Survival* 22(2): 166–172.

Takeyh, R., 2009. *Guardians of the Revolution: Iran and the World in the Age of the Ayatollahs*. Oxford: Oxford University Press.

Taliaferro, Jeffrey W. 2006. "State Building for Future Wars: Neoclassical Realism and the Resource-Extractive State." *Security Studies* 15(3): 464–495.

Telegrams about Dominican Republic; Includes Attachments. 1961. United States. Deputy Special Assistant to the President for National Security Affairs. Secret, Memorandum. August 4, 1961: 33 pp. DNSA collection: CIA Covert Operations III.

Thiem, Alrik. 2013. "Clearly Crisp, and Not Fuzzy: A Reassessment of the (Putative) Pitfalls of Multi-value QCA." *Field Methods* 25(2), 197–207.

Thomas, Scott M. 2001. "Faith, History and Martin Wight: The Role of Religion in the Historical Sociology of the English School of International Relations." *International Affairs* 77(4) (October): 905–929.

Thompson, Hunter S. 2005. *Hey Rube: Blood Sport, the Bush Doctrine, and the Downward Spiral of Dumbness*. New York, NY: Simon & Schuster.

Tiernay, Michael. 2015. "Killing Kony: Leadership Change and Civil War Termination." *Journal of Conflict Resolution* 59(2): 175–206.

Tilly, Charles. *Democracy.* Cambridge: Cambridge University Press, 2007.

Tirman, John. 2009. "Diplomacy, Terrorism, and National Narratives in the United States–Iran relationship." *Critical Studies on Terrorism* 2(3): 527–539.

Trefzger, Douglas W. 2002. "Guatemala's 1952 Agrarian Reform Law: A Critical Reassessment." *International Social Science Review* 77(1–2) (January): 32.

Treier, Shawn and Simon Jackman. 2008. "Democracy as a Latent Variable." *American Journal of Political Science* 52(1): 201–217.

Treverton, Gregory F. "Covert Action and Open Society," *Foreign Policy* 65(5) (Summer): 995–1014.

Treverton, Gregory T., Seth G. Jones, Steven Boraz, and Phillip Lipscy. 2006. *Toward a Theory of Intelligence: Workshop Report.* Washington, DC: RAND Corporation.

Trier, Shawn, and Simon Jackman. 2008. "Democracy as a Latent Variable." *American Journal of Political Science* 52 (January): 201–17.

Ulfelder, Jay. 2005. "Contentious Collective Action and the Breakdown of Authoritarian Regimes." *International Political Science Review* 26(3): 311–334.

Ulfelder, Jay, and Michael Lustik. 2007. "Modeling Transitions to and from Democracy." *Democratization* 14 (April): 351–87.

US Congress, Senate, Select Committee to Study Governmental Operations with Respect to Intelligence Activities, Intelligence Activities and the Rights of Americans, 94 Cong., 2nd sess., 1976, 1.

US Senate. 1975. Report on Alleged Assassination Plots Involving Foreign Leaders: An Interim Report, Select Committee to Study Governmental Operations with respect to Intelligence Activities (the 'Church Committee'). Washington, DC: 94th Congress, 1st Session, Report No. 94-4654, November 1975.

_____. 1961–1962. Foreign Relations of the United States (FRUS). vol. 10, Cuba 1961–1962, and vol. 11, The Cuban Missile Crisis and its Aftermath. Washington DC: USGPO.

Vandenbroucke, Lucien S. 1984. "Anatomy of a Failure: The Decision to Land at the Bay of Pigs." *Political Science Quarterly* 99(3) (Fall): 471–91.

_____. 1993. *Perilous Options: Special Operations as an Instrument of U.S. Foreign Policy.* New York: Oxford University Press.

Van Evera, Stephen. 1999. *Causes of War: Power and the Roots of Conflict*. Ithaca, NY: Cornell University Press.

Van Lierde, Jean. 1972. *Lumumba Speaks: The Speeches and Writings of Patrice Lumumba, 1958–1961.* Boston, MA: Little Brown and Company.

Varshney, Ashutosh. 1999. “Why Democracy Survives.” *Journal of Democracy* 9(3) (January): 36–50.

Verweijen, Judith. 2017. “Strange Battlefield Fellows: The Diagonal Interoperability Between Blue Helmets and the Congolese Army.” *International Peacekeeping* 24(3): 363–387.

Vickers, Adrian. 2005. *A History of Modern Indonesia.* Cambridge, UK: Cambridge University Press.

Vink, Maarten P., and van Vliet, Olaf. 2009. “Not Quite Crisp, Not Yet Fuzzy? Assessing the Potentials and Pitfalls of Multi-value QCA.” *Field Methods* 21(3), 265–289.

_____. 2013. “Potentials and Pitfalls of Multi-value QCA: Response to Them.” *Field Methods* 25(2), 208–213.

Vreeland, James. 2008. “The Effect of Political Regime on Civil War: Unpacking Anocracy.” *Journal of Conflict Resolution* 52 (June): 401–24.

Walt, Stephen M. 1998. “International Relations: One World, Many Theories.” *Foreign Policy* 110 (Spring): 29–46.

Waltz, Kenneth. 1979. *Theory of International Politics*. Long Grove, IL: Waveland Press, Inc.

_____. 1986. “Reflections on Theory of International Politics: A Response to My Critics.” In *Neorealism and Its Critics*, edited by Robert Keohane, 322–345. New York: Columbia University Press.

_____. 2000. “Structural Realism after the Cold War.” *International Security* 25(1): 4–41.

_____. 2001. *Man, the State, and War: A Theoretical Analysis* 3rd Ed. New York: Columbia University Press.

Ward, Michael D. and Andreas Beger. 2017. “Lessons from Near Real-Time Forecasting of Irregular Leadership Changes.” *Journal of Peace Research* 54(2): 141–156.

Warner, Michael, ed. 2001. *Central Intelligence: Origin and Evolution.* Washington, DC: Center for the Study of Intelligence.

Washington, D.C. Office of the Historian, U.S. Department of State, n.d. https://history.state.gov/historicaldocuments/frus1961-63v10/d84. 7 Jan. 2017.

Watson, David, and Lee Anne Clark. 1997. “Extraversion and its Positive Emotional Core.” In *Handbook of Personality Psychology,* edited by R. Hogan, J. A. Johnson, and S. R. Briggs. San Diego, CA: Academic Press.

Weeks, Jessica L. 2012. “Strongmen and Straw Men: Authoritarian Regimes and the Initiation of International Conflict.” *American Political Science Review* 106(2): 326–347.

_____. 2014. *Dictators at War and Peace*. Ithaca, NY: Cornell University Press.

Wedge, Bryant. 1969. "The Case Study of Student Political Violence: Brazil, 1964, and Dominican Republic, 1965." *World Politics* 21(2) (January): 183–206.

Weiner, Tim. 2007. *Legacy of Ashes: The History of the CIA.* New York: Anchor Books.

Weissman, Stephen R. 2010. "An Extraordinary Rendition," *Intelligence and National Security* 25(2): 198–222.

Welch, Claude E. 1970. *Soldier and State in Africa.* Evanston, IL: Northwestern University Press.

Weld, Kristen. 2014. *Paper Cadavers: The Archives of Dictatorship in Guatemala.* Durham, NC: Duke University Press.

Wendt, Alexander. 1999. *Social Theory of International Politics*. Cambridge: Cambridge University Press.

Westerfield, H. Bradford. 1995. *Inside the CIA's Private World: Declassified Articles from the Agency's Internal Journal, 1955–1992.* New Haven, CT: Yale University Press.

Wheeler, Nicholas J. and Timothy Dunne. 1996. "Hedley Bull's Pluralism of the Intellect and Solidarism of the Will," *International Affairs* 72(1) (January): 91–107.

Whitehead, Lawrence. 1999. "Geography and Democratic Destiny." *Journal of Democracy* 10(1) (1999): 74–79.

_____. 2002. *Democratization: Theory and Experience.* Oxford, UK: Oxford University Press.

Wight, Colin. 2006. *Agents, Structures and International Relations: Politics as Ontology*. Cambridge: Cambridge University Press.

Wilber, Donald N. 2000. *Clandestine Service History: Overthrow of Premier Mossadeq of Iran, November 1952–August 1953,* CS Historical Paper No. 208, and March 1954, published October 1969 (Declassified Historical Review Program).

Williams, Gary. 2007. *US-Grenada Relations: Revolution and Intervention in the Backyard.* New York, NY: Palgrave Macmillan.

Winn, Peter. 2005. "Salvador Allende: His Political Life . . . and Afterlife." *Socialism and Democracy* 19–3 (November): 129–159.

Wise, David and Thomas B. Ross. 1965. *The Invisible Government.* New York: Vintage Books.

Wittner, Lawrence S. 1982. *American Intervention in Greece, 1943–1949.* New York: Columbia University Press.

Wolcott, H. 1990. "On Seeking—and Rejecting—Validity in Qualitative Research." In E. W. Eisner and A. Peshkin, (Eds.), *Qualitative Inquiry in Education: The Continuing Debate*. New York: Teachers College Press.

Wolcott, H. 1994. *Transforming Qualitative Data: Description Analysis and Interpretation*. Thousand Oaks, CA: Sage.

Wood, Kristen N. 2014. "Covert Action: A Systems Approach." Master's thesis, Naval Postgraduate School.

Wood R. M. and Gibney M. 2010. "The Political Terror Scale (PTS): A Re-introduction and a Comparison to CIRI." *Human Rights Quarterly* 32(2): 367–400.

Wright, Robert. 2012. "Why Regime Change Won't Work in Iran." *The Atlantic Blog,* January 18, 2012. Accessed September 25, 2016. http://www.theatlantic.com/international/archive/2012/01/why-regime-change-wont-work-iniran/251603/.

Zeilig, Leo. 2008. *Patrice Lumumba: Africa's Lost Leader*. London, UK: Hause Publishing.

Index

Allende, Salvador, 7, 13, 105–113, 121, 128, 130, 132
Arbenz, Jacobo, x, 10, 12, 33–47, 53, 105, 121, 130
assassination, 11–13, 27, 37–38, 56, 63, 66, 75, 78–90, 93, 96, 100, 102, 104, 109, 112, 115, 122

Bay of Pigs invasion, 33, 84, 131
big five traits to leadership model:
  agreeableness, 21, 23, 38, 40–41, 53, 65, 75, 86, 97, 108, 130;
  conscientiousness, 23, 38, 41, 54, 65, 75, 87, 98, 108, 130;
  extraversion, 6, 21, 38, 40, 63, 73, 85, 97, 107, 130;
  neuroticism, 21, 38, 51, 63, 73, 85, 96, 106, 130;
  openness to experience, 6, 22, 38, 40, 52, 64, 74, 86, 97, 107, 130

case studies, vii, xi, 1, 4, 6–7, 9, 11, 14, 31, 48, 68, 79, 93, 103, 106, 115–116, 122–123, 125, 127, 129–133, 136
Castro, Fidel, 7, 12, 69, 83–91, 106, 121–122, 128–129, 132
Chile, vii, 1, 7–8, 13, 90, 105–113, 122, 127–130, 132
Church Committee, ix, 69
CIA, viii, ix, 1, 6, 12, 15, 17, 29, 34–39, 43, 45–47, 56–57, 61, 66, 69, 78–79, 84–90, 102, 109–113, 123–124, 126–127, 137
Cold War, vii, viii–xi, 4, 12, 43–45, 50, 55–56, 100, 102, 116, 119, 124, 126, 134, 137
colonialism, 7, 35, 49, 54–55, 73, 75, 80, 94, 104, 115–116, 128
communism, 31–34, 36, 43, 46–47, 55–56, 58, 70, 72, 80, 89, 91–92, 96, 103–104, 110, 112–113, 131, 133
coup d'état, 12–13
covert action, vii–xi, 1–2, 4–6, 8–17, 28, 31–33, 42, 46–48, 55, 57–59, 61, 67, 70, 80–81, 83–84, 93, 105, 111–113, 115–116, 118–119, 122–127, 129–130, 133, 135–137;
  types of covert action, 125
Cuba, vii, 1, 7–8, 12, 41, 61, 83–92, 106, 121–122, 128–129, 131–132

degree of membership, 2, 3, 8, 11, 17, 32, 58, 80–81, 83, 93, 105, 129
democracy, 4, 7, 12, 17, 19, 21, 25, 30–33, 38, 42, 44–47, 50, 53, 57–58, 74, 76, 78, 92, 94–95, 104–106, 108,

112–113, 116, 119–120, 128–129, 131;
consensus model, 119;
guided democracy, 47, 50, 57–58, 129;
presidentialism, 120
Democratic Republic of the Congo, vii, 1, 7–8, 71–81, 121, 127–128, 130, 132
Diem, Ngo Dinh, 12–13, 93, 95–104, 121, 132
Dominican Republic, vii, ix, 1, 8, 12, 61–70, 128, 132

foreign policy, viii–x, 20, 32, 43, 47, 50, 67–68, 70, 80, 83, 91, 100, 104, 113, 117, 122, 124–125, 137
fuzzy-set qualitative comparative analysis (fsQCA), vii, 11, 58, 126, 133, 135

global instability, vii, ix, 3, 7–8, 14, 18–19, 26, 31–33, 43, 45–46, 48, 55, 57–59, 66, 68–70, 76, 80–81, 89, 91–92, 100, 103–104, 109, 112–113, 119, 124, 126–131, 133–136
Guatemala, vii, x, 1, 7–8, 12, 33–47, 53, 57, 84, 90, 105–106, 118, 121, 127–132

Indonesia, vii, x, 1, 7–8, 12, 40, 47–59, 121–122, 127–132
internal conflict, vii, 3–4, 7–8, 14, 18–19, 27–29, 31–32, 34–35, 43, 45–46, 48, 55, 57–59, 67, 69–70, 77, 79–81, 83, 90–93, 101, 103–104, 110, 113, 119, 124, 126–131, 133–137
Iran, vii, x, 1, 7–8, 12, 17–32, 34, 36, 41, 43, 45, 47, 53, 57, 118, 127–128, 130–132

leader:
characteristics, vii, ix, 3, 6, 8, 14, 17–19, 31–32, 34, 36, 38, 45–46, 48, 50, 57–58, 62, 68–70, 72, 79–81, 83–84, 91–93, 95, 103–106, 112–113, 119, 122, 124–130, 133–137;
survivability, vii, 9, 80, 115
leadership style, vii, 2–3, 9, 11, 13, 15, 80, 113, 118, 132
Lumumba, Patrice, 12, 71–80, 121–122

methodology, 1–7, 9, 11, 13, 15, 126, 128
Mossadegh, Mohammad, 12, 17–26, 29–32, 41, 43, 47, 53, 121, 130

national security, x, 6, 72, 123, 125, 127
neoclassical realism, 1, 14–15, 117, 126
neorealism, 116–117

personality trait, vii, 2, 6, 11, 15, 45, 64–65, 80, 107–108, 118
PKI (Indonesian Communist Party), 47, 56–57
power, viii–ix, 3, 9–11, 15–16, 18–21, 26, 29, 31–32, 34–36, 38, 43–48, 51, 53, 55–56, 58, 61, 63–71, 77, 79–81, 85–88, 91, 93–95, 99–100, 103, 109–110, 113, 115–118, 123, 125–126;
Westminster model, 119
public dissent, vii, ix, 3, 6, 8, 14, 18–19, 24, 31–32, 42, 46, 48, 54, 57–59, 66, 68–70, 75, 80–81, 83, 88, 91–93, 98, 103–104, 108–109, 113, 119, 124, 126–131, 133–137

regime type, vii, 3–4, 7–8, 14, 18, 30–32, 44, 46, 48, 57–59, 67, 69–70, 77, 79–81, 91–92, 101, 103–104, 111, 113, 119, 124, 126–137
Republic of Vietnam, vii, 1, 7–8, 12, 93–104 122, 128–129, 131–132
ruler, vii, 13, 25, 30, 42;
overthrow of ruler, vii, 29, 33–34, 61, 103, 105, 112, 118;
rival, vii, 7, 44, 56, 109, 111, 116, 127

Sukarno, x, 12, 47–59, 121, 129
survival, vii, ix–xi, 1–2, 5–6, 8–11, 13–15, 18, 31, 34, 45, 59, 79–81, 90, 94, 115, 117–120, 122, 124–127, 129–130, 133, 136–137

Trujillo, Rafael, 12, 61–70, 121, 132

variable, vii, 2, 7, 9–11, 13–14, 31–32, 46, 57–58, 70, 81, 92, 104, 113, 118, 126, 128–129, 133;
  dependent, vii, 2, 13–14, 133;
  independent, vii, 11, 13–14, 81
Vietcong, 98, 101, 103

Wayang, 50, 54–55, 57–58

About the Author

Dr. Joy S. Patton was born in Bangkok, Thailand, and lived in Jeddah, Saudi Arabia, as well as Tehran, Iran, prior to the 1979 Revolution. She is a twenty-two-year veteran of the United States military and is an avid traveler and fitness enthusiast. She is an adjunct professor of intelligence studies at American Military University and a credentialed leadership coach through the International Coach Federation.

www.ingramcontent.com/pod-product-compliance
Ingram Content Group UK Ltd.
Pitfield, Milton Keynes, MK11 3LW, UK
UKHW041838031026
15587UKWH00002B/857

9 781793 641717